Introduction to Relativity

Introduction to Relativity

Juliette Backer

WILLFORD PRESS

www.willfordpress.com

Published by Willford Press,
118-35 Queens Blvd., Suite 400,
Forest Hills, NY 11375, USA

ISBN: 978-1-68285-918-6

Cataloging-in-Publication Data

Introduction to relativity / Juliette Backer.
    p. cm.
Includes bibliographical references and index.
ISBN 978-1-68285-918-6
1. Relativity (Physics). 2. Gravitation. 3. Nonrelativistic quantum mechanics.
4. Space and time. 5. Physics. I. Backer, Juliette.
QC173.55 .I58 2020
530.11--dc23

For information on all Willford Press publications
visit our website at www.willfordpress.com

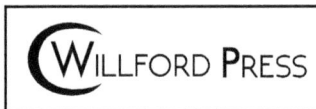

WILLFORD PRESS

TABLE OF CONTENTS

PREFACE

It is with great pleasure that I present this book. It has been carefully written after numerous discussions with my peers and other practitioners of the field. I would like to take this opportunity to thank my family and friends who have been extremely supporting at every step in my life.

Special relativity and general relativity are the two interrelated theories that are usually encompassed in the theory of relativity. Special relativity is applied to all physical phenomena in the absence of gravity. The law of gravitation and its relation to other forces of nature is explained by general relativity. It is applied to cosmological and astrophysical realm such as astronomy. The theory of relativity explores the concepts like spacetime as a unified entity of space and time, kinematic and gravitational time dilation, length contraction and relativity of simultaneity, etc. Relativistic effects are important practical engineering concerns. Satellite-based measurements use relativistic effects to study the motion of an Earth-bound user. This book provides comprehensive insights into the field of relativity. It aims to shed light on some of the unexplored aspects of this field. The book is appropriate for those seeking detailed information in this area.

The chapters below are organized to facilitate a comprehensive understanding of the subject:

Chapter – Introduction

The theories of relativity serve as the basis to understand the geometry of the universe as well as the various cosmic processes. There are two major theories within this area of research, namely, general relativity and special relativity. The topics elaborated in this chapter will help in gaining a better perspective about various aspects of relativity.

Chapter – Special Relativity

The special theory of relativity establishes the relationship between time and space for problems that classical physics is unable to understand or explain. Some of the focus areas of special relativity are speed of light, rapidity, proper length and relativistic mass. The chapter closely examines these key concepts of special relativity to provide an extensive understanding of the subject.

Chapter – Phenomena of Special Relativity

Some of the phenomena and concepts that are studied within special relativity are mass–energy equivalence, time dilation, length contraction, relativity of simultaneity, relativistic Doppler effect, Thomas precession, ladder paradox, twin paradox, etc. This chapter discusses in detail these phenomena and concepts related to special relativity.

Chapter – General Relativity

The theory of general relativity states that the observed gravitational effect between masses results from their warping of spacetime. Some of the areas which are studied in general relativity include equivalence principle, Penrose diagram, geodesics in general relativity, Mach's principle, linearized gravity, Raychaudhuri equation, etc. The diverse areas of general relativity have been thoroughly discussed in this chapter.

Chapter – Phenomena of General Relativity

Some of the common phenomena which are studied within general relativity are black hole, event horizon, frame-dragging, gravitational singularity, gravitational time dilation, gravitational redshift, Shapiro time delay, gravitational wave and gravitational lensing. This chapter has been carefully written to provide an easy understanding of these phenomena of general relativity.

Juliette Backer

Introduction

The theories of relativity serve as the basis to understand the geometry of the universe as well as the various cosmic processes. There are two major theories within this area of research, namely, general relativity and special relativity. The topics elaborated in this chapter will help in gaining a better perspective about various aspects of relativity.

Relativity

Relativity is a set of wide-ranging physical theories formed by the German-born physicist Albert Einstein. With his theories of special relativity (1905) and general relativity (1915), Einstein overthrew many assumptions underlying earlier physical theories, redefining in the process the fundamental concepts of space, time, matter, energy, and gravity. Along with quantum mechanics, relativity is central to modern physics. In particular, relativity provides the basis for understanding cosmic processes and the geometry of the universe itself.

"Special relativity" is limited to objects that are moving with respect to inertial frames of reference—i.e, in a state of uniform motion with respect to one another such that an observer cannot, by purely mechanical experiments, distinguish one from the other. Beginning with the behaviour of light (and all other electromagnetic radiation), the theory of special relativity draws conclusions that are contrary to everyday experience but fully confirmed by experiments. Special relativity revealed that the speed of light is a limit that can be approached but not reached by any material object; it is the origin of the most famous equation in science, $E = mc^2$; and it has led to other tantalizing outcomes, such as the "twin paradox".

"General relativity" is concerned with gravity, one of the fundamental forces in the universe. (The others are electricity and magnetism, which have been unified as electromagnetism, the strong force, and the weak force.) Gravity defines macroscopic behaviour, and so general relativity describes large-scale physical phenomena such as planetary dynamics, the birth and death of stars, black holes, and the evolution of the universe.

Special and general relativity have profoundly affected physical science and human existence, most dramatically in applications of nuclear energy and nuclear weapons. Additionally, relativity and its rethinking of the fundamental categories of space and time have provided a basis for certain philosophical, social, and artistic interpretations that have influenced human culture in different ways.

Cosmology before Relativity

The Mechanical Universe

Relativity changed the scientific conception of the universe, which began in efforts to grasp the dynamic behaviour of matter. In Renaissance times, the great Italian physicist Galileo Galilei moved beyond Aristotle's philosophy to introduce the modern study of mechanics, which requires quantitative measurements of bodies moving in space and time. His work and that of others led to basic concepts, such as velocity, which is the distance a body covers in a given direction per unit time; acceleration, the rate of change of velocity; mass, the amount of material in a body; and force, a push or pull on a body.

The next major stride occurred in the late 17th century, when the British scientific genius Isaac Newton formulated his three famous laws of motion, the first and second of which are of special concern in relativity. Newton's first law, known as the law of inertia, states that a body that is not acted upon by external forces undergoes no acceleration—either remaining at rest or continuing to move in a straight line at constant speed. Newton's second law states that a force applied to a body changes its velocity by producing an acceleration that is proportional to the force and inversely proportional to the mass of the body. In constructing his system, Newton also defined space and time, taking both to be absolutes that are unaffected by anything external. Time, he wrote, "flows equally," while space "remains always similar and immovable."

Newton's laws proved valid in every application, as in calculating the behaviour of falling bodies, but they also provided the framework for his landmark law of gravity. Beginning with the (perhaps mythical) observation of a falling apple and then considering the Moon as it orbits Earth, Newton concluded that an invisible force acts between the Sun and its planets. He formulated a comparatively simple mathematical expression for the gravitational force; it states that every object in the universe attracts every other object with a force that operates through empty space and that varies with the masses of the objects and the distance between them.

The law of gravity was brilliantly successful in explaining the mechanism behind Kepler's laws of planetary motion, which the German astronomer Johannes Kepler had formulated at the beginning of the 17th century. Newton's mechanics and law of gravity, along with his assumptions about the nature of space and time, seemed wholly successful in explaining the dynamics of the universe, from motion on Earth to cosmic events.

Light and the Ether

However, this success at explaining natural phenomena came to be tested from an unexpected direction—the behaviour of light, whose intangible nature had puzzled philosophers and scientists for centuries. In 1865 the Scottish physicist James Clerk Maxwellshowed that light is an electromagnetic wave with oscillating electrical and magnetic components. Maxwell's equations predicted that electromagnetic waves would travel through empty space at a speed of almost exactly 3×10^8 metres per second (186,000 miles per second)—i.e., according with the measured speed of light. Experiments soon confirmed the electromagnetic nature of light and established its speed as a fundamental parameter of the universe.

Maxwell's remarkable result answered long-standing questions about light, but it raised another fundamental issue: if light is a moving wave, what medium supports it? Ocean waves and sound waves consist of the progressive oscillatory motion of molecules of water and of atmospheric gases, respectively. But what is it that vibrates to make a moving light wave? Or to put it another way, how does the energy embodied in light travel from point to point?

For Maxwell and other scientists of the time, the answer was that light traveled in a hypothetical medium called the ether (aether). Supposedly, this medium permeated all space without impeding the motion of planets and stars; yet it had to be more rigid than steel so that light waves could move through it at high speed, in the same way that a taut guitar string supports fast mechanical vibrations. Despite this contradiction, the idea of the ether seemed essential—until a definitive experiment disproved it.

In 1887 the German-born American physicist A.A. Michelson and the American chemist Edward Morley made exquisitely precise measurements to determine how Earth's motion through the ether affected the measured speed of light. In classical mechanics, Earth's movement would add to or subtract from the measured speed of light waves, just as the speed of a ship would add to or subtract from the speed of ocean waves as measured from the ship. But the Michelson-Morley experiment had an unexpected outcome, for the measured speed of light remained the same regardless of Earth's motion. This could only mean that the ether had no meaning and that the behaviour of light could not be explained by classical physics. The explanation emerged, instead, from Einstein's theory of special relativity.

Special Relativity

Einstein's Gedankenexperiments

Scientists such as Austrian physicist Ernst Mach and French mathematician Henri Poincaréhad critiqued classical mechanics or contemplated the behaviour of light and the meaning of the ether before Einstein. Their efforts provided a background for Einstein's unique approach to understanding the universe, which he called in his native German a *Gedankenexperiment*, or "thought experiment."

Einstein described how at age 16 he watched himself in his mind's eye as he rode on a light wave and gazed at another light wave moving parallel to his. According to classical physics, Einstein should have seen the second light wave moving at a relative speed of zero. However, Einstein knew that Maxwell's electromagnetic equations absolutely require that light always move at 3×10^8 metres per second in a vacuum. Nothing in the theory allows a light wave to have a speed of zero. Another problem arose as well: if a fixed observer sees light as having a speed of 3×10^8 metres per second, whereas an observer moving at the speed of light sees light as having a speed of zero, it would mean that the laws of electromagnetism depend on the observer. But in classical mechanics the same laws apply for all observers, and Einstein saw no reason why the electromagnetic laws should not be equally universal. The constancy of the speed of light and the universality of the laws of physics for all observers are cornerstones of special relativity.

Starting Points and Postulates

In developing special relativity, Einstein began by accepting what experiment and his own thinking showed to be the true behaviour of light, even when this contradicted classical physics or the usual perceptions about the world.

The fact that the speed of light is the same for all observers is inexplicable in ordinary terms. If a passenger in a train moving at 100 km per hour shoots an arrow in the train's direction of motion at 200 km per hour, a trackside observer would measure the speed of the arrow as the sum of the two speeds, or 300 km per hour. In analogy, if the train moves at the speed of light and a passenger shines a laser in the same direction, then common sense indicates that a trackside observer should see the light moving at the sum of the two speeds, or twice the speed of light (6×10^8 metres per second).

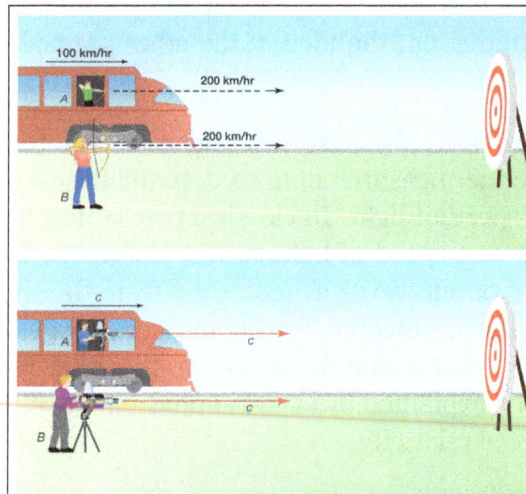

Invariance of the speed of light.

Arrows shot from a moving train (A) and from a stationary location (B) will arrive at a target at different velocities—in this case, 300 and 200 km/hr, respectively, because of the motion of the train. However, such commonsense addition of velocities does not apply to light. Even for a train traveling at the speed of light, both laser beams, A and B, have the same velocity: c.

While such a law of addition of velocities is valid in classical mechanics, the Michelson-Morley experiment showed that light does not obey this law. This contradicts common sense; it implies, for instance, that both a train moving at the speed of light and a light beam emitted from the train arrive at a point farther along the track at the same instant.

Nevertheless, Einstein made the constancy of the speed of light for all observers a postulate of his new theory. As a second postulate, he required that the laws of physics have the same form for all observers. Then Einstein extended his postulates to their logical conclusions to form special relativity.

Experimental Evidence for Special Relativity

Because relativistic changes are small at typical speeds for macroscopic objects, the confirmation of special relativity has relied on either the examination of subatomic bodies at high speeds or the measurement of small changes by sensitive instrumentation. For example, ultra-accurate clocks were placed on a variety of commercial airliners flying at one-millionth the speed of light. After two days of continuous flight, the time shown by the airborne clocks differed by fractions of a microsecond from that shown by a synchronized clock left on Earth, as predicted.

Larger effects are seen with elementary particles moving at speeds close to that of light. One such experiment involved muons, elementary particles created by cosmic rays in Earth's atmosphere at an altitude of about 9 km (30,000 feet). At 99.8 percent of the speed of light, the muons should reach sea level in 31 microseconds, but measurements showed that it took only 2 microseconds. The reason is that, relative to the moving muons, the distance of 9 km contracted to 0.58 km (1,900 feet). Similarly, a relativistic mass increase has been confirmed in measurements on fast-moving elementary particles, where the change is large.

Such results leave no doubt that special relativity correctly describes the universe, although the theory is difficult to accept at a visceral level. Some insight comes from Einstein's comment that in relativity the limiting speed of light plays the role of an infinite speed. At infinite speed, light would traverse any distance in zero time. Similarly, according to the relativistic equations, an observer riding a light wave would see lengths contract to zero and clocks stop ticking as the universe approached him at the speed of light. Effectively, relativity replaces an infinite speed limit with the finite value of 3×10^8 metres per second.

General Relativity

Roots of General Relativity

Because Isaac Newton's law of gravity served so well in explaining the behaviour of the solar system, the question arises why it was necessary to develop a new theory of gravity. The answer is that Newton's theory violates special relativity, for it requires an unspecified "action at a distance" through which any two objects—such as the Sun and Earth—instantaneously pull each other, no matter how far apart. However, instantaneous response would require the gravitational interaction to propagate at infinite speed, which is precluded by special relativity.

In practice, this is no great problem for describing our solar system, for Newton's law gives valid answers for objects moving slowly compared with light. Nevertheless, since Newton's theory cannot be conceptually reconciled with special relativity, Einstein turned to the development of general relativity as a new way to understand gravitation.

Experimental Evidence for General Relativity

Soon after the theory of general relativity was published in 1915, the English astronomer Arthur Eddington considered Einstein's prediction that light rays are bent near a massive body, and he realized that it could be verified by carefully comparing star positions in images of the Sun taken during a solar eclipse with images of the same region of space taken when the Sun was in a different portion of the sky. Verification was delayed by World War I, but in 1919 an excellent opportunity presented itself with an especially long total solar eclipse, in the vicinity of the bright Hyades star cluster, that was visible from northern Brazil to the African coast. Eddington led one expedition to Príncipe, an island off the African coast, and Andrew Crommelin of the Royal Greenwich Observatory led a second expedition to Sobral, Brazil. After carefully comparing photographs from both expeditions with reference photographs of the Hyades, Eddington declared that the starlight had been deflected about 1.75 seconds of arc, as predicted by general relativity. (The same effect produces gravitational lensing, where a massive cosmic object focuses light from another object beyond it to produce a distorted or magnified image. The astronomical discovery of gravitational lenses in 1979 gave additional support for general relativity.)

In 1919, observation of a solar eclipse confirmed Einstein's prediction that light is bent in the presence of mass. This experimental support for his general theory of relativity garnered him instant worldwide acclaim.

Further evidence came from the planet Mercury. In the 19th century, it was found that Mercury does not return to exactly the same spot every time it completes its elliptical orbit. Instead, the ellipse rotates slowly in space, so that on each orbit the perihelion—the point of closest approach to the Sun—moves to a slightly different angle. Newton's law of gravity could not explain this perihelion shift, but general relativity gave the correct orbit.

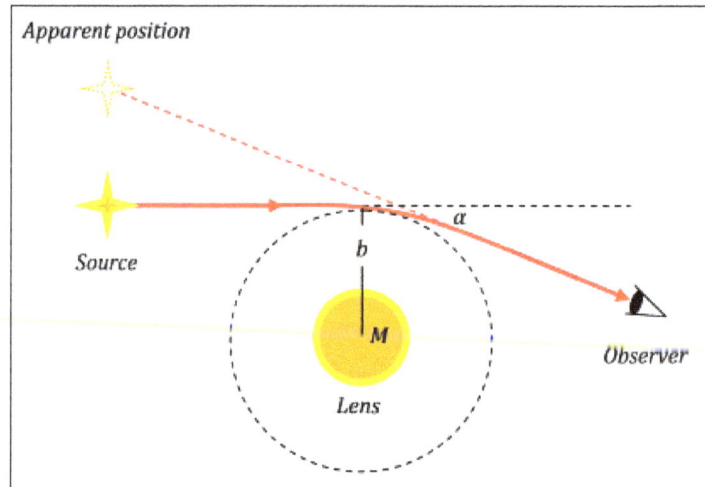

Another confirmed prediction of general relativity is that time dilates in a gravitational field, meaning that clocks run slower as they approach the mass that is producing the field. This has been measured directly and also through the gravitational redshift of light. Time dilation causes light to vibrate at a lower frequency within a gravitational field; thus, the light is shifted toward a longer wavelength—that is, toward the red. Other measurements have verified the equivalence principle by showing that inertial and gravitational mass are precisely the same.

The most striking prediction of general relativity is that of gravitational waves. Electromagnetic waves are caused by accelerated electrical charges and are detected when they put other charges into motion. Similarly, gravitational waves would be caused by masses in motion and are detected when they initiate motion in other masses. However, gravity is very weak compared with electromagnetism. Only a huge cosmic event, such as the collision of two stars, can generate detectable gravitational waves. Efforts to sense gravitational waves began in the 1960s, and such waves were first detected in 2015 when LIGO observed two black holes 1.3 million light-years away spiralling into each other.

Applications of Relativistic Ideas

Although relativistic effects are negligible in ordinary life, relativistic ideas appear in a range of areas from fundamental science to civilian and military technology.

Elementary Particles

The relationship $E = mc^2$ is essential in the study of subatomic particles. It determines the energy required to create particles or to convert one type into another and the energy released when a particle is annihilated. For example, two photons, each of energy E, can collide to form two particles, each with mass $m = E/c^2$. This pair-production process is one step in the early evolution of the universe, as described in the big-bang model.

Particle Accelerators

Knowledge of elementary particles comes primarily from particle accelerators. These machines raise subatomic particles, usually electrons or protons, to nearly the speed of light. When these energetic bullets smash into selected targets, they elucidate how subatomic particles interact and often produce new species of elementary particles.

Particle accelerators could not be properly designed without special relativity. In the type called an electron synchrotron, for instance, electrons gain energy as they traverse a huge circular raceway. At barely below the speed of light, their mass is thousands of times larger than their rest mass. As a result, the magnetic field used to hold the electrons in circular orbits must be thousands of times stronger than if the mass did not change.

Fission and Fusion: Bombs and Stellar Processes

Energy is released in two kinds of nuclear processes. In nuclear fission a heavy nucleus, such as uranium, splits into two lighter nuclei; in nuclear fusion two light nuclei combine into a heavier one. In each process the total final mass is less than the starting mass. The difference appears as energy according to the relation $E = \Delta m c^2$, where Δm is the mass deficit.

Fission is used in atomic bombs and in reactors that produce power for civilian and military applications. The fusion of hydrogen into helium is the energy source in stars and provides the power of a hydrogen bomb. Efforts are now under way to develop controllable hydrogen fusion as a clean, abundant power source.

Global Positioning System

The global positioning system (GPS) depends on relativistic principles. A GPS receiver determines its location on Earth's surface by processing radio signals from four or more satellites. The distance to each satellite is calculated as the product of the speed of light and the time lag between transmission and reception of the signal. However, Earth's gravitational field and the motion of the satellites cause time-dilation effects, and Earth's rotation also has relativistic implications. Hence, GPS technology includes relativistic corrections that enable positions to be calculated to within several centimetres.

Cosmology

Cosmology, the study of the structure and origin of the universe, is intimately connected with gravity, which determines the macroscopic behaviour of all matter. General relativity has played a role in cosmology since the early calculations of Einstein and Friedmann. Since then, the theory has provided a framework for accommodating observational results, such as Hubble's discovery of the expanding universe in 1929, as well as the big-bang model, which is the generally accepted explanation of the origin of the universe.

The latest solutions of Einstein's field equations depend on specific parameters that characterize the fate and shape of the universe. One is Hubble's constant, which defines how rapidly the universe is expanding; the other is the density of matter in the universe, which determines the strength of gravity. Below a certain critical density, gravity would be weak enough that the universe would expand forever, so that space would be unlimited. Above that value, gravity would be strong enough to make the universe shrink back to its original minute size after a finite period of expansion, a process called the "big crunch." In this case, space would be limited or bounded like the surface of a sphere. Current efforts in observational cosmology focus on measuring the most accurate possible values of Hubble's constant and of critical density.

Relativity, Quantum Theory and Unified Theories

Cosmic behaviour on the biggest scale is described by general relativity. Behaviour on the subatomic scale is described by quantum mechanics, which began with the work of the German physicist Max Planck in 1900 and treats energy and other physical quantities in discrete units called quanta. A central goal of physics has been to combine relativity theory and quantum theory into an overarching "theory of everything" describing all physical phenomena. Quantum theory explains electromagnetism and the strong and weak forces, but a quantum description of the remaining fundamental force of gravity has not been achieved.

After Einstein developed relativity, he unsuccessfully sought a so-called unified field theory with a space-time geometry that would encompass all the fundamental forces. Other theorists have attempted to merge general relativity with quantum theory, but the two approaches treat forces in fundamentally different ways. In quantum theory, forces arise from the interchange of certain elementary particles, not from the shape of space-time. Furthermore, quantum effects are thought to cause a serious distortion of space-time at an extremely small scale called the Planck length, which is much smaller than the size of elementary particles. This suggests that quantum gravity cannot be understood without treating space-time at unheard-of scales.

Although the connection between general relativity and quantum mechanics remains elusive, some progress has been made toward a fully unified theory. In the 1960s, the electroweak theory provided partial unification, showing a common basis for electromagnetism and the weak force within quantum theory. Recent research suggests that superstring theory, in which elementary particles are represented not as mathematical points but as extremely small strings vibrating in 10 or more dimensions, shows promise for supporting complete unification, including gravitation. However, until confirmed by experimental results, superstring theory will remain an untested hypothesis.

Intellectual and Cultural Impact of Relativity

Reactions in General Culture

The impact of relativity has not been limited to science. Special relativity arrived on the scene at the beginning of the 20th century, and general relativity became widely known after World War I—eras when a new sensibility of "modernism" was becoming defined in art and literature. In addition, the confirmation of general relativity provided by the solar eclipse of 1919 received wide publicity. Einstein's 1921 Nobel Prize for Physics (awarded for his work on the photon nature of light), as well as the popular perception that relativity was so complex that few could grasp it, quickly turned Einstein and his theories into cultural icons.

The ideas of relativity were widely applied—and misapplied—soon after their advent. Some thinkers interpreted the theory as meaning simply that all things are relative, and they employed this concept in arenas distant from physics. The Spanish humanist philosopher and essayist José Ortega y Gasset, for instance, wrote in The Modern Theme:

> "The theory of Einstein is a marvelous proof of the harmonious multiplicity of all possible points of view. If the idea is extended to morals and aesthetics, we shall come to experience history and life in a new way."

The revolutionary aspect of Einstein's thought was also seized upon, as by the American art critic Thomas Craven, who in 1921 compared the break between classical and modern art to the break between Newtonian and Einsteinian ideas about space and time.

Some saw specific relations between relativity and art arising from the idea of a four-dimensional space-time continuum. In the 19th century, developments in geometry led to popular interest in a fourth spatial dimension, imagined as somehow lying at right angles to all three of the ordinary dimensions of length, width, and height. Edwin Abbott's Flatland was the first popular presentation of these ideas. Other works of fantasy that followed spoke of the fourth dimension as an arena apart from ordinary existence.

Einstein's four-dimensional universe, with three spatial dimensions and one of time, is conceptually different from four spatial dimensions. But the two kinds of four-dimensional world became conflated in interpreting the new art of the 20th century. Early Cubist works by Pablo Picasso that simultaneously portrayed all sides of their subjects became connected with the idea of higher dimensions in space, which some writers attempted to relate to relativity. In 1949, for example, the art historian Paul LaPorte wrote that "the new pictorial idiom created by [C]ubism is most satisfactorily explained by applying to it the concept of the space-time continuum." Einstein specifically rejected this view, saying, "This new artistic 'language' has nothing in common with the Theory of Relativity." Nevertheless, some artists explicitly explored Einstein's ideas. In the new Soviet Union of the 1920s, for example, the poet and illustrator Vladimir Mayakovsky, a founder of the artistic movement called Russian Futurism, or Suprematism, hired an expert to explain relativity to him.

The widespread general interest in relativity was reflected in the number of books written to elucidate the subject for nonexperts. Einstein's popular exposition of special and general relativity appeared almost immediately, in 1916, other scientists, such as the Russian mathematician Aleksandr Friedmann and the British astronomer Arthur Eddington, wrote popular books on the subjects in the 1920s. Such books continued to appear decades later.

When relativity was first announced, the public was typically awestruck by its complexity, a justified response to the intricate mathematics of general relativity. But the abstract, nonvisceral nature of the theory also generated reactions against its apparent violation of common sense. These reactions included a political undertone; in some quarters, it was considered undemocratic to present or support a theory that could not be immediately understood by the common person.

In contemporary usage, general culture has accepted the ideas of relativity—the impossibility of faster-than-light travel, $E = mc^2$, time dilation and the twin paradox, the expanding universe, and black holes and wormholes—to the point where they are immediately recognized in the media and provide plot devices for works of science fiction. Some of these ideas have gained meaning beyond their strictly scientific ones; in the business world, for instance, "black hole" can mean an unrecoverable financial drain.

Philosophical Considerations

In 1925 the British philosopher Bertrand Russell, in his ABC of Relativity, suggested that Einstein's work would lead to new philosophical concepts. Relativity has indeed had a great effect on philosophy, illuminating some issues that go back to the ancient Greeks. The idea of the ether, invoked in the late 19th century to carry light waves, harks back to Aristotle. He divided the world into earth, air, fire, and water, with the ether (aether) as the fifth element representing the pure celestial sphere. The Michelson-Morley experiment and relativity eliminated the last vestiges of this idea.

Relativity also changed the meaning of geometry as it was developed in Euclid's Elements (c. 300 BCE). Euclid's system relied on the axiom "a straight line is the shortest distance between two points," among others that seemed self-evidently true. Straight lines also played a special role in Euclid's Optics as the paths followed by light rays. To philosophers such as the German Immanuel Kant, Euclid's straight-line axiom represented a deep level of truth. But general relativity makes it possible scientifically to examine space like any other physical quantity—that is, to investigate Euclid's premises. It is now known that space-time is curved near stars; no straight lines exist there, and light follows curved geodesics. Like Newton's law of gravity, Euclid's geometry correctly describes reality under certain conditions, but its axioms are not absolutely fundamental and universal, for the cosmos includes non-Euclidean geometries as well.

Considering its scientific breadth, its recasting of people's view of reality, its ability to describe the entire universe, and its influence outside science, Einstein's relativity stands among the most significant and influential of scientific theories.

Special Relativity 2

- **Special Theory of Relativity**
- **Frame of Reference**
- **Lorentz Transformation**
- **Speed of Light**
- **Michelson–Morley Experiment**
- **Rapidity**
- **Maxwell's Equations**
- **Proper Length**
- **Proper Time**
- **Relativistic Mass**

The special theory of relativity establishes the relationship between time and space for problems that classical physics is unable to understand or explain. Some of the focus areas of special relativity are speed of light, rapidity, proper length and relativistic mass. The chapter closely examines these key concepts of special relativity to provide an extensive understanding of the subject.

Special Theory of Relativity

The special theory of relativity establishes the relationship between space and time for problems that classical physics (Newtonian physics) fail to understand or explain.

The Postulates

The whole premise upon which special relativity is built consists of two simple postulates:

1. The laws of physics hold in non-accelerating frames of reference.

2. The speed of light in a vacuum remains constant for all observers.

The reason these postulates were framed was that Einstein found an inconsistency when he tried to apply Newtonian Mechanics to Maxwell's equations of electromagnetism. This led Einstein to believe that Newtonian Mechanics did not work according to Maxwell's Equations of Electromagnetism. However, it was not possible for the speed of time to be a variable, as it had been rigorously analyzed and experimented upon by numerous scientists. Although Newtonian physics were highly instrumental when it came to everyday calculations, they failed miserably when considering high-velocity physics (physics at the speed of light).

The Mathematical Foundation

The funny thing about the special theory of relativity was that the math required for it had been invented before the theory itself. Albert Einstein went through many mathematical tools before finally settling on the mathematical framework of Hendrik Lorentz. Hendrik Lorentz was a Dutch physicist who invented the Lorentz transformation, which helped accommodate the notion that multiple frames of reference can be calculated simultaneously (the foundation upon which the special theory of relativity was written). Using the infamous equations of Lorentz Transformation, Albert Einstein published the scientific paper "On the electrodynamics of moving bodies", which we today know as the Special Theory of Relativity. The equations are as follows:

$$\gamma = 1 / \left(1 - \left(c^2 / v^2\right)\right)^{0.5}$$

$$ct' = ct - \beta * x$$

$$x' = x - \beta * ct$$

Phenomena Explained by Special Relativity

The case point scenarios where special theory comes into play can be proven through Lorentz transformation. Special cases where the physics calculations must be performed at velocities close to the speed of light are where special relativity comes to the fore.

This is when two events appear to occur at two different locations simultaneously in the reference frame of one observer, but may also appear to happen non-simultaneously in the frame of another observer. To take an example, let's consider a man standing on a moving platform. There are two light sources equidistant from him. Now, let another man be present at a different frame of reference than the man on the platform. Whether the platform moves relative to the man or the other way around, both observers do not report the same timing of the light sources flashing! The man on the train always reports that both light beams are produced simultaneously, but the man observing from the platform reports one light beam being faster than the other (the one closest to him). How can this be and who is wrong?

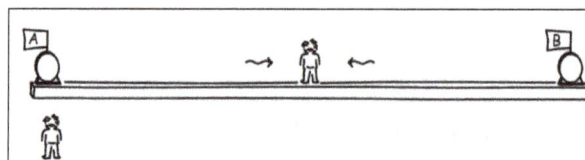

The observer moving relative to the platform.

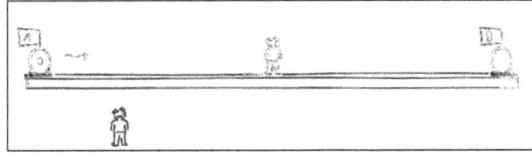

The platform moving relative to the observer.

The answer to this question is that neither observer is wrong. The two observers disagree on the timing of the event and the synchronicity with which the events occur. This is because both observers view the same event from different spatial orientations to the relative motion of the occurring event. Therefore, the same event's occurrence appears differently for different observers due to the relative motion being observed separately. This is relativity of simultaneity.

The second situation where this theory comes into play is length contraction. This is a very surprising phenomenon arguing that the measurement of an object by one observer might be highly different from that of another observer. The simplest way to explain this is with another "moving platform" example. If you were standing on Earth and a spaceship were to pass above you at 10% the speed of light you would be able to see the spaceship and note its length. Now, if you saw the same spaceship pass you a second time at 85% the speed of light, you would see a significantly shorter spaceship. At 99% the speed of light, you would probably see nothing but a line.

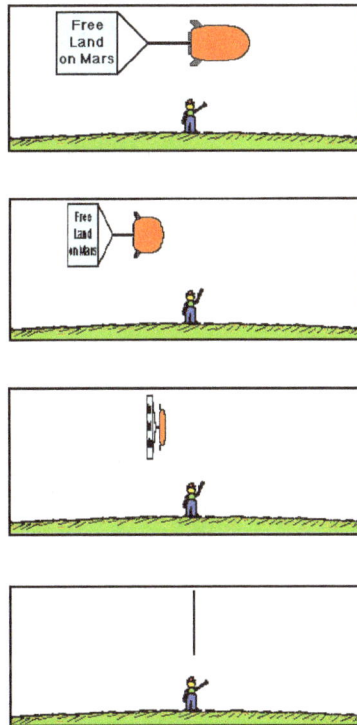

One of the final points we will touch on is causality and the prohibition of motion faster than the speed of light. This is by far the most interesting principle that we can deduce from the special theory of relativity. The meaning of causality in classical physics implies that no action can occur before its cause. In Einstein's theory of special relativity, causality means that an effect cannot occur from a cause that is not in the back (past) light cone of that event.

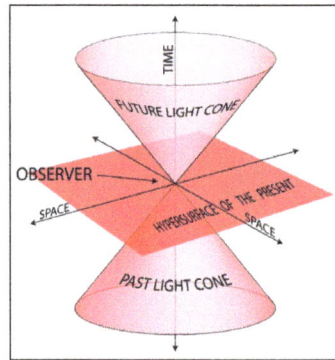

A light cone is the path that a flash of light takes when emanating from a single point in space and traveling in all directions of spacetime. If you imagine the current event (or the present moment) to be a plane horizon, then the light from that spreads out as a growing circle of events, resulting in a cone. This light cone that is formed is known as the future light cone. This light cone is a mathematical representation of all future events that will occur after the event horizon. The inverse of this future light cone is what we call the past light cone. If anything were to cross the speed of light, numerous paradoxes would begin to occur. One famous hypothesis is that if one were to travel faster than the speed of light, there is a high chance that instead of going into the future light cone, one would be able to break the space-time barrier and move into the past light cone. Therefore, most scientists feel that it will be easier to travel into the future than into the past.

There are other phenomena that can be explained with the help the of the special theory of relativity, such as time dilation, the composition of velocities and certain optical effects. It may have taken a while for the public to understand or accept Einstein's wild postulations, but over time, we have found that his scientific theories – and his genius-level reputation – continue to be supported and validated.

Frame of Reference

In physics, a frame of reference (or reference frame) consists of an abstract coordinate system and the set of physical reference points that uniquely fix (locate and orient) the coordinate system and standardize measurements.

In n dimensions, n + 1 reference points are sufficient to fully define a reference frame. Using rectangular (Cartesian) coordinates, a reference frame may be defined with a reference point at the origin and a reference point at one unit distance along each of the n coordinate axes.

In Einsteinian relativity, reference frames are used to specify the relationship between a moving observer and the phenomenon or phenomena under observation. In this context, the phrase often becomes "observational frame of reference" (or "observational reference frame"), which implies that the observer is at rest in the frame, although not necessarily located at its origin. A relativistic reference frame includes (or implies) the coordinate time, which does not correspond across different frames moving relatively to each other. The situation thus differs from Galilean relativity, where all possible coordinate times are essentially equivalent.

Different aspects of "Frame of Reference"

The need to distinguish between the various meanings of "frame of reference" has led to a variety of terms. For example, sometimes the type of coordinate system is attached as a modifier, as in Cartesian frame of reference. Sometimes the state of motion is emphasized, as in rotating frame of reference. Sometimes the way it transforms to frames considered as related is emphasized as in Galilean frame of reference. Sometimes frames are distinguished by the scale of their observations, as in macroscopic and microscopic frames of reference.

The term observational frame of reference is used when emphasis is upon the state of motion rather than upon the coordinate choice or the character of the observations or observational apparatus. In this sense, an observational frame of reference allows study of the effect of motion upon an entire family of coordinate systems that could be attached to this frame. On the other hand, a coordinate system may be employed for many purposes where the state of motion is not the primary concern. For example, a coordinate system may be adopted to take advantage of the symmetry of a system. In a still broader perspective, the formulation of many problems in physics employs generalized coordinates, normal modes or eigenvectors, which are only indirectly related to space and time. We therefore take observational frames of reference, coordinate systems, and observational equipment as independent concepts, separated as below:

- An observational frame (such as an inertial frame or non-inertial frame of reference) is a physical concept related to state of motion.

- A coordinate system is a mathematical concept, amounting to a choice of language used to describe observations. Consequently, an observer in an observational frame of reference can choose to employ any coordinate system (Cartesian, polar, curvilinear, generalized) to describe observations made from that frame of reference. A change in the choice of this coordinate system does not change an observer's state of motion, and so does not entail a change in the observer's *observational* frame of reference. This viewpoint can be found elsewhere as well. Which is not to dispute that some coordinate systems may be a better choice for some observations than are others.

- Choice of what to measure and with what observational apparatus is a matter separate from the observer's state of motion and choice of coordinate system.

Here is a quotation applicable to moving observational frames $\Re$ and various associated Euclidean three-space coordinate systems [R, R', etc.]:

> "We first introduce the notion of reference frame, itself related to the idea of observer: the reference frame is, in some sense, the "Euclidean space carried by the observer". Let us give a more mathematical definition: the reference frame is the set of all points in the Euclidean space with the rigid body motion of the observer. The frame, denoted $\Re$, is said to move with the observer. The spatial positions of particles are labelled relative to a frame $\Re$ by establishing a coordinate system R with origin O. The corresponding set of axes, sharing the rigid body motion of the frame $\Re$, can be considered to give a physical realization of $\Re$. In a frame $\Re$, coordinates are changed from R to R' by carrying out, at each instant of time, the same coordinate transformation on the components of intrinsic objects (vectors and tensors) introduced to represent physical quantities in this frame".

And this on the utility of separating the notions of $\Re$ and [R, R', etc.]:

> "As noted by Brillouin, a distinction between mathematical sets of coordinates and physical frames of reference must be made. The ignorance of such distinction is the source of much confusion the dependent functions such as velocity for example, are measured with respect to a physical reference frame, but one is free to choose any mathematical coordinate system in which the equations are specified".

And this, also on the distinction between $\Re$ and [R, R', etc.]:

> "The idea of a reference frame is really quite different from that of a coordinate system. Frames differ just when they define different spaces (sets of rest points) or times (sets of simultaneous events). So the ideas of a space, a time, of rest and simultaneity, go inextricably together with that of frame. However, a mere shift of origin, or a purely spatial rotation of space coordinates results in a new coordinate system. So frames correspond at best to classes of coordinate systems".

And from J. D. Norton:

> "In traditional developments of special and general relativity it has been customary not to distinguish between two quite distinct ideas. The first is the notion of a coordinate system, understood simply as the smooth, invertible assignment of four numbers to events in spacetime neighborhoods. The second, the frame of reference, refers to an idealized system used to assign such numbers To avoid unnecessary restrictions, we can divorce this arrangement from metrical notions. Of special importance for our purposes is that each frame of reference has a definite state of motion at each event of spacetime. Within the context of special relativity and as long as we restrict ourselves to frames of reference in inertial motion, then little of importance depends on the difference between an inertial frame of reference and the inertial coordinate system it induces. This comfortable circumstance ceases immediately once we begin to consider frames of reference in nonuniform motion even within special relativity. More recently, to negotiate the obvious ambiguities of Einstein's treatment, the notion of frame of reference has reappeared as a structure distinct from a coordinate system".

Extension to coordinate systems using generalized coordinates underlies the Hamiltonian and Lagrangian formulations of quantum field theory, classical relativistic mechanics, and quantum gravity.

Coordinate Systems

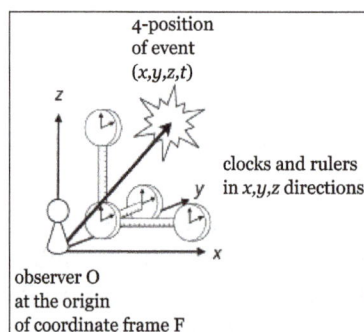

An observer O, situated at the origin of a local set of coordinates – a frame of reference F. The observer in this frame uses the coordinates (x, y, z, t) to describe a spacetime event, shown as a star.

Although the term "coordinate system" is often used (particularly by physicists) in a nontechnical sense, the term "coordinate system" does have a precise meaning in mathematics, and sometimes that is what the physicist means as well.

A coordinate system in mathematics is a facet of geometry or of algebra, in particular, a property of manifolds (for example, in physics, configuration spaces or phase spaces). The coordinates of a point r in an n-dimensional space are simply an ordered set of n numbers:

$$\mathbf{r} = [x^1, x^2, \ldots, x^n].$$

In a general Banach space, these numbers could be (for example) coefficients in a functional expansion like a Fourier series. In a physical problem, they could be spacetime coordinates or normal mode amplitudes. In a robot design, they could be angles of relative rotations, linear displacements, or deformations of joints. Here we will suppose these coordinates can be related to a Cartesian coordinate system by a set of functions:

$$x^j = x^j(x, y, z, \ldots), \quad j = 1, \ldots, n,$$

where x, y, z, etc. are the n Cartesian coordinates of the point. Given these functions, coordinate surfaces are defined by the relations:

$$x^j(x, y, z, \ldots) = \text{constant}, \quad j = 1, \ldots, n.$$

The intersection of these surfaces define coordinate lines. At any selected point, tangents to the intersecting coordinate lines at that point define a set of basis vectors $\{e_1, e_2, \ldots, e_n\}$ at that point. That is:

$$e_i(\mathbf{r}) = \lim_{\epsilon \to 0} \frac{\mathbf{r}\left(x^1, \ldots, x^i + \epsilon, \ldots, x^n\right) - \mathbf{r}\left(x^1, \ldots, x^i, \ldots, x^n\right)}{\epsilon}, \quad i = 1, \ldots, n,$$

which can be normalized to be of unit length.

Coordinate surfaces, coordinate lines, and basis vectors are components of a coordinate system. If the basis vectors are orthogonal at every point, the coordinate system is an orthogonal coordinate system.

An important aspect of a coordinate system is its metric tensor g_{ik}, which determines the arc length ds in the coordinate system in terms of its coordinates:

$$(ds)^2 = g_{ik} \, dx^i \, dx^k,$$

where repeated indices are summed over.

As is apparent from these remarks, a coordinate system is a mathematical construct, part of an axiomatic system. There is no necessary connection between coordinate systems and physical motion (or any other aspect of reality). However, coordinate systems can include time as a coordinate, and can be used to describe motion. Thus, Lorentz transformations and Galilean transformations may be viewed as coordinate transformations.

Observational Frames of Reference

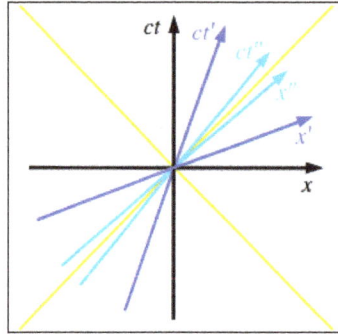

Three frames of reference in special relativity. The black frame is at rest. The primed frame moves at 40% of light speed, and the double primed frame at 80%. Note the scissors-like change as speed increases.

An observational frame of reference, often referred to as a *physical frame of reference*, a *frame of reference*, or simply a *frame*, is a physical concept related to an observer and the observer's state of motion. Here we adopt the view expressed by Kumar and Barve: an observational frame of reference is characterized *only by its state of motion*. However, there is lack of unanimity on this point. In special relativity, the distinction is sometimes made between an *observer* and a *frame*. According to this view, a *frame* is an *observer* plus a coordinate lattice constructed to be an orthonormal right-handed set of spacelike vectors perpendicular to a timelike vector. This restricted view is not used here, and is not universally adopted even in discussions of relativity. In general relativity the use of general coordinate systems is common.

There are two types of observational reference frame: inertial and non-inertial. An inertial frame of reference is defined as one in which all laws of physics take on their simplest form. In special relativity these frames are related by Lorentz transformations, which are parametrized by rapidity. In Newtonian mechanics, a more restricted definition requires only that Newton's first law holds true; that is, a Newtonian inertial frame is one in which a free particle travels in a straight line at constant speed, or is at rest. These frames are related by Galilean transformations. These relativistic and Newtonian transformations are expressed in spaces of general dimension in terms of representations of the Poincaré group and of the Galilean group.

In contrast to the inertial frame, a non-inertial frame of reference is one in which fictitious forces must be invoked to explain observations. An example is an observational frame of reference centered at a point on the Earth's surface. This frame of reference orbits around the center of the Earth, which introduces the fictitious forces known as the Coriolis force, centrifugal force, and gravitational force. (All of these forces including gravity disappear in a truly inertial reference frame, which is one of free-fall.)

Measurement Apparatus

A further aspect of a frame of reference is the role of the measurement apparatus (for example, clocks and rods) attached to the frame. This question and is of particular interest in quantum mechanics, where the relation between observer and measurement is still under discussion.

In physics experiments, the frame of reference in which the laboratory measurement devices are at rest is usually referred to as the laboratory frame or simply "lab frame." An example would be the frame in which the detectors for a particle accelerator are at rest. The lab frame in some experiments is an inertial frame, but it is not required to be (for example the laboratory on the surface of the Earth in many physics experiments is not inertial). In particle physics experiments, it is often useful to transform energies and momenta of particles from the lab frame where they are measured, to the center of momentum frame "COM frame" in which calculations are sometimes simplified, since potentially all kinetic energy still present in the COM frame may be used for making new particles.

In this connection it may be noted that the clocks and rods often used to describe observers' measurement equipment in thought, in practice are replaced by a much more complicated and indirect metrology that is connected to the nature of the vacuum, and uses atomic clocks that operate according to the standard model and that must be corrected for gravitational time dilation.

In fact, Einstein felt that clocks and rods were merely expedient measuring devices and they should be replaced by more fundamental entities based upon, for example, atoms and molecules.

Types

- Body-fixed frames of reference
- Space-fixed frames of reference
- Inertial frames of reference
- Non-Inertial frames of reference

Examples of Inertial Frames of Reference

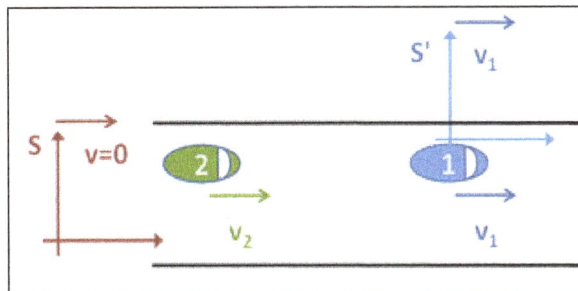

Two cars moving at different but constant velocities observed from stationary inertial frame
S attached to the road and moving inertial frame S'attached to the first car.

Consider a situation common in everyday life. Two cars travel along a road, both moving at constant velocities. At some particular moment, they are separated by 200 metres. The car in front is travelling at 22 metres per second and the car behind is travelling at 30 metres per second. If we want to find out how long it will take the second car to catch up with the first, there are three obvious "frames of reference" that we could choose.

First, we could observe the two cars from the side of the road. We define our "frame of reference" S as follows. We stand on the side of the road and start a stop-clock at the exact moment that the second car passes us, which happens to be when they are a distance $d = 200$ m apart. Since neither of the cars is accelerating, we can determine their positions by the following formulas, where $x_1(t)$ is the position in meters of car one after time t in seconds and $x_2(t)$ is the position of car two after time t.

$$x_1(t) = d + v_1 t = 200 + 22t, \quad x_2(t) = v_2 t = 30t.$$

Notice that these formulas predict at $t = 0$ s the first car is 200 m down the road and the second car is right beside us, as expected. We want to find the time at which $x_1 = x_2$. Therefore, we set $x_1 = x_2$ and solve for t, that is:

$$200 + 22t = 30t,$$
$$8t = 200,$$
$$t = 25 \text{ seconds}.$$

Alternatively, we could choose a frame of reference S' situated in the first car. In this case, the first car is stationary and the second car is approaching from behind at a speed of $v_2 - v_1 = 8$ m/s. In order to catch up to the first car, it will take a time of $\dfrac{d}{v_2 - v_1} = \dfrac{200}{8}$ s, that is, 25 seconds, as before.

Note how much easier the problem becomes by choosing a suitable frame of reference. The third possible frame of reference would be attached to the second car. That example resembles the case just discussed, except the second car is stationary and the first car moves backward towards it at 8 m/s.

It would have been possible to choose a rotating, accelerating frame of reference, moving in a complicated manner, but this would have served to complicate the problem unnecessarily. It is also necessary to note that one is able to convert measurements made in one coordinate system to another. For example, suppose that your watch is running five minutes fast compared to the local standard time. If you know that this is the case, when somebody asks you what time it is, you are able to deduct five minutes from the time displayed on your watch in order to obtain the correct time. The measurements that an observer makes about a system depend therefore on the observer's frame of reference (you might say that the bus arrived at 5 past three, when in fact it arrived at three).

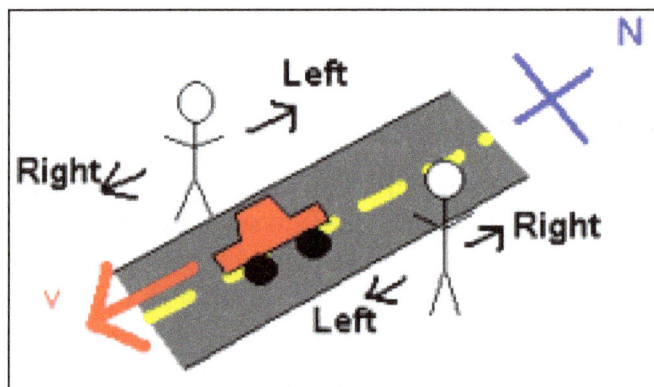

Simple-minded frame-of-reference example.

For a simple example involving only the orientation of two observers, consider two people standing, facing each other on either side of a north-south street. A car drives past them heading south.

For the person facing east, the car was moving towards the right. However, for the person facing west, the car was moving toward the left. This discrepancy is because the two people used two different frames of reference from which to investigate this system.

For a more complex example involving observers in relative motion, consider Alfred, who is standing on the side of a road watching a car drive past him from left to right. In his frame of reference, Alfred defines the spot where he is standing as the origin, the road as the x-axis and the direction in front of him as the positive y-axis. To him, the car moves along the x axis with some velocity v in the positive x-direction. Alfred's frame of reference is considered an inertial frame of reference because he is not accelerating (ignoring effects such as Earth's rotation and gravity).

Now consider Betsy, the person driving the car. Betsy, in choosing her frame of reference, defines her location as the origin, the direction to her right as the positive x-axis, and the direction in front of her as the positive y-axis. In this frame of reference, it is Betsy who is stationary and the world around her that is moving – for instance, as she drives past Alfred, she observes him moving with velocity v in the negative y-direction. If she is driving north, then north is the positive y-direction; if she turns east, east becomes the positive y-direction.

Finally, as an example of non-inertial observers, assume Candace is accelerating her car. As she passes by him, Alfred measures her acceleration and finds it to be a in the negative x-direction. Assuming Candace's acceleration is constant, what acceleration does Betsy measure? If Betsy's velocity v is constant, she is in an inertial frame of reference, and she will find the acceleration to be the same as Alfred in her frame of reference, a in the negative y-direction. However, if she is accelerating at rate A in the negative y-direction (in other words, slowing down), she will find Candace's acceleration to be $a' = a - A$ in the negative y-direction—a smaller value than Alfred has measured. Similarly, if she is accelerating at rate A in the positive y-direction (speeding up), she will observe Candace's acceleration as $a' = a + A$ in the negative y-direction—a larger value than Alfred's measurement.

Frames of reference are especially important in special relativity, because when a frame of reference is moving at some significant fraction of the speed of light, then the flow of time in that frame does not necessarily apply in another frame. The speed of light is considered to be the only true constant between moving frames of reference.

It is important to note some assumptions made above about the various inertial frames of reference. Newton, for instance, employed universal time, as explained by the following example. Suppose that you own two clocks, which both tick at exactly the same rate. You synchronize them so that they both display exactly the same time. The two clocks are now separated and one clock is on a fast moving train, traveling at constant velocity towards the other. According to Newton, these two clocks will still tick at the same rate and will both show the same time. Newton says that the rate of time as measured in one frame of reference should be the same as the rate of time in another. That is, there exists a "universal" time and all other times in all other frames of reference will run at the same rate as this universal time irrespective of their position and velocity. This concept of time and simultaneity was later generalized by Einstein in his special theory of relativity where he developed transformations between inertial frames of reference based upon the universal nature of physical laws and their economy of expression (Lorentz transformations).

The definition of inertial reference frame can also be extended beyond three-dimensional Euclidean space. Newton's assumed a Euclidean space, but general relativity uses a more general geometry. As an example of why this is important, consider the geometry of an ellipsoid. In this geometry, a "free" particle is defined as one at rest or traveling at constant speed on a geodesic path. Two free particles may begin at the same point on the surface, traveling with the same constant speed in different directions. After a length of time, the two particles collide at the opposite side of the ellipsoid. Both "free" particles traveled with a constant speed, satisfying the definition that no forces were acting. No acceleration occurred and so Newton's first law held true. This means that the particles were in inertial frames of reference. Since no forces were acting, it was the geometry of the situation which caused the two particles to meet each other again. In a similar way, it is now common to describe that we exist in a four-dimensional geometry known as spacetime. In this picture, the curvature of this 4D space is responsible for the way in which two bodies with mass are drawn together even if no forces are acting. This curvature of spacetime replaces the force known as gravity in Newtonian mechanics and special relativity.

Non-inertial Frames

Here the relation between inertial and non-inertial observational frames of reference is considered. The basic difference between these frames is the need in non-inertial frames for fictitious forces, as described below.

An accelerated frame of reference is often delineated as being the "primed" frame, and all variables that are dependent on that frame are notated with primes, e.g. x', y', a'.

The vector from the origin of an inertial reference frame to the origin of an accelerated reference frame is commonly notated as R. Given a point of interest that exists in both frames, the vector from the inertial origin to the point is called r, and the vector from the accelerated origin to the point is called r'. From the geometry of the situation, we get:

$$\mathbf{r} = \mathbf{R} + \mathbf{r}'.$$

Taking the first and second derivatives of this with respect to time, we obtain:

$$\mathbf{v} = \mathbf{V} + \mathbf{v}',$$
$$\mathbf{a} = \mathbf{A} + \mathbf{a}'.$$

where V and A are the velocity and acceleration of the accelerated system with respect to the inertial system and v and a are the velocity and acceleration of the point of interest with respect to the inertial frame.

These equations allow transformations between the two coordinate systems; for example, we can now write Newton's second law as:

$$\mathbf{F} = m\mathbf{a} = m\mathbf{A} + m\mathbf{a}'.$$

When there is accelerated motion due to a force being exerted there is manifestation of inertia. If an electric car designed to recharge its battery system when decelerating is switched to braking, the batteries are recharged, illustrating the physical strength of manifestation of inertia. However, the manifestation of inertia does not prevent acceleration (or deceleration), for manifestation of inertia occurs in response to change in velocity due to a force. Seen from the perspective of a rotating frame of reference the manifestation of inertia appears to exert a force (either in centrifugal direction, or in a direction orthogonal to an object's motion, the Coriolis effect).

A common sort of accelerated reference frame is a frame that is both rotating and translating (an example is a frame of reference attached to a CD which is playing while the player is carried). This arrangement leads to the equation:

$$\mathbf{a} = \mathbf{a}' + \dot{\omega} \times \mathbf{r}' + 2\omega \times \mathbf{v}' + \omega \times (\omega \times \mathbf{r}') + \mathbf{A}_0,$$

or, to solve for the acceleration in the accelerated frame,

$$\mathbf{a}' = \mathbf{a} - \dot{\omega} \times \mathbf{r}' - 2\omega \times \mathbf{v}' - \omega \times (\omega \times \mathbf{r}') - \mathbf{A}_0.$$

Multiplying through by the mass m gives:

$$\mathbf{F}' = \mathbf{F}_{physical} + \mathbf{F}'_{Euler} + \mathbf{F}'_{Coriolis} + \mathbf{F}'_{centripetal} - m\mathbf{A}_0,$$

Where:

$$\mathbf{F}'_{Euler} = -m\dot{\omega} \times \mathbf{r}' \text{ (Euler force)},$$

$$\mathbf{F}'_{Coriolis} = -2m\omega \times \mathbf{v}' \text{ (Coriolis force)},$$

$$\mathbf{F}'_{centrifugal} = -m\omega \times (\omega \times \mathbf{r}') = m(\omega^2 \mathbf{r}' - (\omega \cdot \mathbf{r}')\omega)(\text{centrifugal force}).$$

Lorentz Transformation

Lorentz Transform are needed to relate events observed in inertial frames moving at relative speeds v approaching the speed of light c.

We begin with the assumption that the same physical laws are valid in all inertial frames. Consider two inertial coordinate frames S and the "primed" frame S' moving with a constant velocity v along their common x-axis. The planes $y = 0$ and $z = 0$ always coincide with the planes $y' = 0$ and $z' = 0$, and we can set the zero points on the clocks so that $t = t' = 0$ at the instant when $x = x'$.

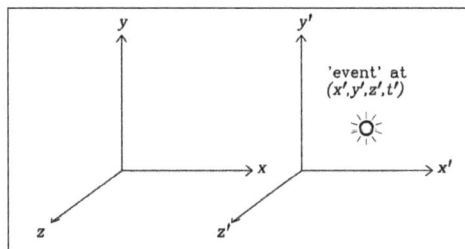

Thus (x, y, z, t) might be coordinates in the rest frame of the Galaxy and (x', y', z', t') an inertial frame moving with the instantaneous velocity of a (non-inertial) cosmic-ray electron. An *event* is something that occurs at one point (x', y', z', t') in space and time, such as firing the flash on a camera at rest in the S' frame.

According to the "intuitively obvious" Galilean relativity, the event coordinates in the unprimed frame S are:

$$x = x' + vt' \quad y = y' \quad z = z' \quad t = t'$$

Unfortunately, our intuition is shaped by experiences with objects moving at speeds much less than the speed of light in a vacuum, and it does not correspond to reality at speeds approaching c. We "know" by experience that time is absolute; that is, $t = t'$ in all inertial frames. Galilean relativity implies that parallel velocities simply add. The speed of a photon emitted by the flash in the x' direction will be seen as:

$$c_x = \frac{dx}{dt} = \frac{d(x' + vt')}{dt'} = \frac{dx'}{dt'} + v = c_x + v$$

by an observer in the frame S. Thus Galilean relativity is inconsistent with both observation and Maxwell's equations, which correctly predict that the speed of light in a vacuum is the same for all observers in all inertial frames $(c_x = c_x)$, regardless of their relative velocities v. We therefore drop the assumption that time is absolute.

The Lorentz transform is the only coordinate transform consistent with both relativity (equivalence of inertial frames) and the existance of some still-unspecified invariant speed (for example, the speed of light, or even ∞). Note that the latter assumption is much weaker than assuming $t = t'$ because if the invariant speed is infinite, then the assumption has no consequences and the Lorentz transform reduces to the Galilean transform. The Lorentz transform can be derived with two reasonable assumptions. We will assume that space is both *homogeneous* and *isotropic*: that is, the laws of physics do not change from one place to another, or with orientation of the coordinate frames.

Isotropy implies that the observers moving with the two frames agree on their relative speed $|v|$ since a 180 degree coordinate rotation exchanges the roles of the two frames. That rotation should have no effect if space is isotropic. The assumption of homogeneity implies that any transformation from one inertial frame to another is *linear*; e.g., $y' = Ay + B$ where A and B are constants. Any nonlinear terms (e.g., $y' = Ay + B + Cy^2$) would cause the transform itself to vary under coordinate translations and C must therefore be zero in a homogeneous space.

Since we can choose coordinate frames so $y = 0$ when $y' = 0$, linearity requires that $y' = Ay$, where A is some constant scale factor. Reversing the coordinate directions by 180 degrees reverses the roles of S and , so $y = A'y'$. Only $A = A' = \pm 1$ is consistent with isotropy. We can reject the negative solution $A = -1$ because it implies $y = -y'$ when $v = 0$, so the Lorentz transforms for the y and z coordinates are:

$$y = y' \quad \text{and likewise} \quad z = z'$$

in agreement with the Galilean transform.

We can proceed in a similar way with the x-coordinate. Linearity requires:

$$x = \gamma'(x' + vt') \text{ and } x' = \gamma(x - vt)$$

where γ' and γ are (still unspecified) constant scale factors. Reversing the directions of S and S' gives:

$$x = \gamma'(x' - vt') \text{ and } x' = \gamma(x + vt).$$

Reversing the roles of the two frames gives:

$$x = \gamma(x' - vt') \text{ and } x' = \gamma(x + vt).$$

These two pairs of equations imply $\gamma = \gamma'$; that is, the observers in S and   also agree on the scale factor γ associated with the relative velocity v.

Suppose that there is some speed c which is the same in all inertial frames. (We already know from Maxwell's equations and by experiment that c is the speed of light in a vacuum, but for this argument, it could be any speed, even $c = \infty$, in which case the Lorentz transform would end up being identical to the Galilean transform.) Then $x = ct$ implies $x' = ct'$ and:

$$ct = \gamma t'(c + v) \text{ and } ct' = \gamma t(c - v)$$

The product of these these two equations is:

$$c^2 tt' = \gamma^2 tt'(c + v)(c - v).$$

We solve for γ, which is called the *Lorentz factor*:

$$\gamma = \left(1 - \frac{v^2}{c^2}\right)^{-1/2}$$

Again, the negative solution to this equation can be rejected as unphysical. The x-coordinate transform is:

$$x = \gamma(x' + vt') \text{ and } x' = \gamma(x - vt)$$

Eliminating x from this pair of equations yields:

$$t = \gamma(t' + vx'/c^2)$$

The Lorentz transform of special relativity is thus:

$$x = \gamma(x' + vt') \quad y = y' \quad z = z' \quad t = \gamma(t' + \beta x'/c)$$
$$x' = \gamma(x' + vt') \quad y' = y \quad z' = z \quad t' = \gamma(t + \beta x/c)$$

Where $\equiv v/c$. Note that the Galilean transform is just the limit of the Lorentz transform as $c \to \infty$. If $(\Delta x, \Delta y, \Delta z, \Delta t)$ and $(\Delta x', \Delta y', \Delta z', \Delta t')$ are the coordinate differences between two events, the differential form of the (linear) Lorentz transform is:

$$\Delta x = \gamma\left(\Delta x' + v\Delta t'\right) \quad \Delta y = \Delta y' \quad \Delta z = \Delta z' \quad \Delta t = \gamma\left(\Delta t' + \beta\Delta x'/c\right)$$

$$\Delta x' = \gamma\left(\Delta x + v\Delta t\right) \quad \Delta y' = \Delta y \quad \Delta z' = \Delta z \quad \Delta t' = \gamma\left(\Delta t + \beta\Delta x/c\right)$$

Speed of Light

The speed of light in vacuum, commonly denoted c, is a universal physical constant important in many areas of physics. Its exact value is 299,792,458 metres per second (approximately 300,000 km/s (186,000 mi/s)). It is exact because by international agreement a metre is defined as the length of the path travelled by light in vacuum during a time interval of 1/299792458 second. According to special relativity, c is the upper limit for the speed at which conventional matter and information can travel. Though this speed is most commonly associated with light, it is also the speed at which all massless particles and field perturbations travel in vacuum, including electromagnetic radiation and gravitational waves. Such particles and waves travel at c regardless of the motion of the source or the inertial reference frame of the observer. Particles with nonzero rest mass can approach c, but can never actually reach it. In the special and general theories of relativity, c interrelates space and time, and also appears in the famous equation of mass–energy equivalence $E = mc^2$.

The speed at which light propagates through transparent materials, such as glass or air, is less than c; similarly, the speed of electromagnetic waves in wire cables is slower than c. The ratio between c and the speed v at which light travels in a material is called the refractive index n of the material ($n = c / v$). For example, for visible light the refractive index of glass is typically around 1.5, meaning that light in glass travels at $c / 1.5 \approx 200,000$ km/s (124,000 mi/s); the refractive index of air for visible light is about 1.0003, so the speed of light in air is about 299,700 km/s (186,220 mi/s), which is about 90 km/s (56 mi/s) slower than c.

For many practical purposes, light and other electromagnetic waves will appear to propagate instantaneously, but for long distances and very sensitive measurements, their finite speed has noticeable effects. In communicating with distant space probes, it can take minutes to hours for a message to get from Earth to the spacecraft, or vice versa. The light seen from stars left them many years ago, allowing the study of the history of the universe by looking at distant objects. The finite speed of light also limits the theoretical maximum speed of computers, since information must be sent within the computer from chip to chip. The speed of light can be used with time of flight measurements to measure large distances to high precision.

Ole Rømer first demonstrated in 1676 that light travels at a finite speed (as opposed to instantaneously) by studying the apparent motion of Jupiter's moon Io. In 1865, James Clerk Maxwell proposed that light was an electromagnetic wave, and therefore travelled at the speed c appearing in his theory of electromagnetism. In 1905, Albert Einstein postulated that the speed of light c with respect to any inertial frame is a constant and is independent of the motion of the light source. He explored the consequences of that postulate by deriving the theory of relativity and in doing so showed that the parameter c had relevance outside of the context of light and electromagnetism.

After centuries of increasingly precise measurements, in 1975 the speed of light was known to be 299792458 m/s (983571056 ft/s; 186282.397 mi/s) with a measurement uncertainty of 4 parts per billion. In 1983, the metre was redefined in the International System of Units (SI) as the distance travelled by light in vacuum in 1/299792458 of a second.

Numerical Value, Notation and Units

The speed of light in vacuum is usually denoted by a lowercase c, for "constant" or the Latin *celeritas*. In 1856, Wilhelm Eduard Weber and Rudolf Kohlrausch had used c for a different constant later shown to equal $\sqrt{2}$ times the speed of light in vacuum. Historically, the symbol V was used as an alternative symbol for the speed of light, introduced by James Clerk Maxwell in 1865. In 1894, Paul Drude redefined c with its modern meaning. Einstein used V in his original German-language papers on special relativity in 1905, but in 1907 he switched to c, which by then had become the standard symbol for the speed of light.

Sometimes c is used for the speed of waves in *any* material medium, and c_0 for the speed of light in vacuum. This subscripted notation, which is endorsed in official SI literature, has the same form as other related constants: namely, μ_0 for the vacuum permeability or magnetic constant, ε_0 for the vacuum permittivity or electric constant, and Z_0 for the impedance of free space.

Since 1983, the metre has been defined in the International System of Units (SI) as the distance light travels in vacuum in $\frac{1}{299792458}$ of a second. This definition fixes the speed of light in vacuum at exactly 299,792,458 m/s. As a dimensional physical constant, the numerical value of c is different for different unit systems. In branches of physics in which c appears often, such as in relativity, it is common to use systems of natural units of measurement or the geometrized unit system where $c = 1$. Using these units, c does not appear explicitly because multiplication or division by 1 does not affect the result.

Fundamental Role in Physics

The speed at which light waves propagate in vacuum is independent both of the motion of the wave source and of the inertial frame of reference of the observer. This invariance of the speed of light was postulated by Einstein in 1905, after being motivated by Maxwell's theory of electromagnetism and the lack of evidence for the luminiferous aether; it has since been consistently confirmed by many experiments. It is only possible to verify experimentally that the two-way speed of light (for example, from a source to a mirror and back again) is frame-independent, because it is impossible to measure the one-way speed of light (for example, from a source to a distant detector) without some convention as to how clocks at the source and at the detector should be synchronized. However, by adopting Einstein synchronization for the clocks, the one-way speed of light becomes equal to the two-way speed of light by definition. The special theory of relativity explores the consequences of this invariance of c with the assumption that the laws of physics are the same in all inertial frames of reference. One consequence is that c is the speed at which all massless particles and waves, including light, must travel in vacuum.

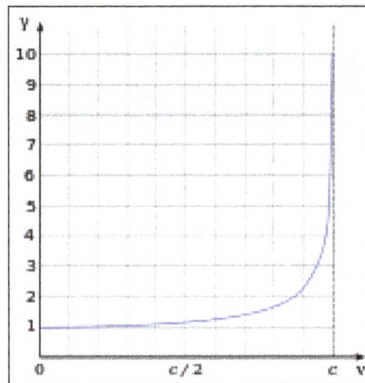

The Lorentz factor γ as a function of velocity. It starts at 1 and approaches infinity as v approaches c.

Special relativity has many counterintuitive and experimentally verified implications. These include the equivalence of mass and energy ($E = mc^2$), length contraction (moving objects shorten), and time dilation (moving clocks run more slowly). The factor γ by which lengths contract and times dilate is known as the Lorentz factor and is given by $\gamma = (1 - v^2/c^2)^{-1/2}$, where v is the speed of the object. The difference of γ from 1 is negligible for speeds much slower than c, such as most everyday speeds—in which case special relativity is closely approximated by Galilean relativity—but it increases at relativistic speeds and diverges to infinity as v approaches c. For example, a time dilation factor of $\gamma = 2$ occurs at a relative velocity of 86.6% of the speed of light ($v = .866c$). Similarly, a time dilation factor of $\gamma = 10$ occurs at $v = 99.5\%\ c$.

The results of special relativity can be summarized by treating space and time as a unified structure known as spacetime (with c relating the units of space and time), and requiring that physical theories satisfy a special symmetry called Lorentz invariance, whose mathematical formulation contains the parameter c. Lorentz invariance is an almost universal assumption for modern physical theories, such as quantum electrodynamics, quantum chromodynamics, the Standard Model of particle physics, and general relativity. As such, the parameter c is ubiquitous in modern physics, appearing in many contexts that are unrelated to light. For example, general relativity predicts that c is also the speed of gravity and of gravitational waves. In non-inertial frames of reference (gravitationally curved spacetime or accelerated reference frames), the *local* speed of light is constant and equal to c, but the speed of light along a trajectory of finite length can differ from c, depending on how distances and times are defined.

It is generally assumed that fundamental constants such as c have the same value throughout spacetime, meaning that they do not depend on location and do not vary with time. However, it has been suggested in various theories that the speed of light may have changed over time. No conclusive evidence for such changes has been found, but they remain the subject of ongoing research.

It also is generally assumed that the speed of light is isotropic, meaning that it has the same value regardless of the direction in which it is measured. Observations of the emissions from nuclear energy levels as a function of the orientation of the emitting nuclei in a magnetic field, and of rotating optical resonators have put stringent limits on the possible two-way anisotropy.

Upper Limit on Speeds

According to special relativity, the energy of an object with rest mass m and speed v is given by γmc^2, where γ is the Lorentz factor defined above. When v is zero, γ is equal to one, giving rise to the famous $E = mc^2$ formula for mass–energy equivalence. The γ factor approaches infinity as v approaches c, and it would take an infinite amount of energy to accelerate an object with mass to the speed of light. The speed of light is the upper limit for the speeds of objects with positive rest mass, and individual photons cannot travel faster than the speed of light. This is experimentally established in many tests of relativistic energy and momentum.

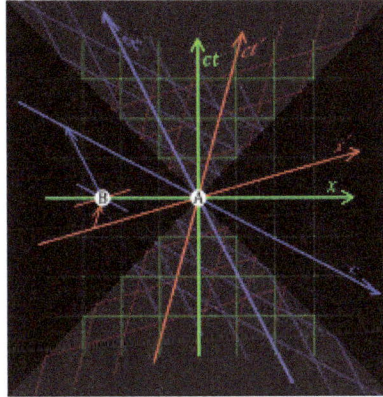

Event A precedes B in the red frame, is simultaneous with B in the green frame, and follows B in the blue frame.

More generally, it is normally impossible for information or energy to travel faster than c. One argument for this follows from the counter-intuitive implication of special relativity known as the relativity of simultaneity. If the spatial distance between two events A and B is greater than the time interval between them multiplied by c then there are frames of reference in which A precedes B, others in which B precedes A, and others in which they are simultaneous. As a result, if something were travelling faster than c relative to an inertial frame of reference, it would be travelling backwards in time relative to another frame, and causality would be violated. In such a frame of reference, an "effect" could be observed before its "cause". Such a violation of causality has never been recorded, and would lead to paradoxes such as the tachyonic antitelephone.

Faster than Light Observations and Experiments

There are situations in which it may seem that matter, energy, or information travels at speeds greater than c, but they do not. For example, as is discussed in the propagation of light in a medium, many wave velocities can exceed c. For example, the phase velocity of X-rays through most glasses can routinely exceed c, but phase velocity does not determine the velocity at which waves convey information.

If a laser beam is swept quickly across a distant object, the spot of light can move faster than c, although the initial movement of the spot is delayed because of the time it takes light to get to the distant object at the speed c. However, the only physical entities that are moving are the laser and its emitted light, which travels at the speed c from the laser to the various positions of the spot. Similarly, a shadow projected onto a distant object can be made to move faster than c, after a delay in time. In neither case does any matter, energy, or information travel faster than light.

The rate of change in the distance between two objects in a frame of reference with respect to which both are moving (their closing speed) may have a value in excess of c. However, this does not represent the speed of any single object as measured in a single inertial frame.

Certain quantum effects appear to be transmitted instantaneously and therefore faster than c, as in the EPR paradox. An example involves the quantum states of two particles that can be entangled. Until either of the particles is observed, they exist in a superposition of two quantum states. If the particles are separated and one particle's quantum state is observed, the other particle's quantum state is determined instantaneously (i.e., faster than light could travel from one particle to the other). However, it is impossible to control which quantum state the first particle will take on when it is observed, so information cannot be transmitted in this manner.

Another quantum effect that predicts the occurrence of faster-than-light speeds is called the Hartman effect: under certain conditions the time needed for a virtual particle to tunnel through a barrier is constant, regardless of the thickness of the barrier. This could result in a virtual particle crossing a large gap faster-than-light. However, no information can be sent using this effect.

So-called superluminal motion is seen in certain astronomical objects, such as the relativistic jets of radio galaxies and quasars. However, these jets are not moving at speeds in excess of the speed of light: the apparent superluminal motion is a projection effect caused by objects moving near the speed of light and approaching Earth at a small angle to the line of sight: since the light which was emitted when the jet was farther away took longer to reach the Earth, the time between two successive observations corresponds to a longer time between the instants at which the light rays were emitted.

In models of the expanding universe, the farther galaxies are from each other, the faster they drift apart. This receding is not due to motion *through* space, but rather to the expansion of space itself. For example, galaxies far away from Earth appear to be moving away from the Earth with a speed proportional to their distances. Beyond a boundary called the Hubble sphere, the rate at which their distance from Earth increases becomes greater than the speed of light.

Propagation of Light

In classical physics, light is described as a type of electromagnetic wave. The classical behaviour of the electromagnetic field is described by Maxwell's equations, which predict that the speed c with which electromagnetic waves (such as light) propagate through the vacuum is related to the distributed capacitance and inductance of the vacuum, otherwise respectively known as the electric constant ε_0 and the magnetic constant μ_0, by the equation,

$$c = \frac{1}{\sqrt{\varepsilon_0 \mu_0}}.$$

In modern quantum physics, the electromagnetic field is described by the theory of quantum electrodynamics (QED). In this theory, light is described by the fundamental excitations (or quanta) of the electromagnetic field, called photons. In QED, photons are massless particles and thus, according to special relativity, they travel at the speed of light in vacuum.

Extensions of QED in which the photon has a mass have been considered. In such a theory, its speed would depend on its frequency, and the invariant speed c of special relativity would then be the upper limit of the speed of light in vacuum. No variation of the speed of light with frequency has been observed in rigorous testing, putting stringent limits on the mass of the photon. The limit obtained depends on the model used: if the massive photon is described by Proca theory, the experimental upper bound for its mass is about 10^{-57} grams; if photon mass is generated by a Higgs mechanism, the experimental upper limit is less sharp, $m \leq 10^{-14}$ eV/c² (roughly 2×10^{-47} g).

Another reason for the speed of light to vary with its frequency would be the failure of special relativity to apply to arbitrarily small scales, as predicted by some proposed theories of quantum gravity. In 2009, the observation of the spectrum of gamma-ray burst GRB 090510 did not find any difference in the speeds of photons of different energies, confirming that Lorentz invariance is verified at least down to the scale of the Planck length ($l_p = \sqrt{\hbar G/c^3} \approx 1.6163 \times 10^{-35}$m) divided by 1.2.

In a Medium

In a medium, light usually does not propagate at a speed equal to c; further, different types of light wave will travel at different speeds. The speed at which the individual crests and troughs of a plane wave (a wave filling the whole space, with only one frequency) propagate is called the phase velocity v_p. An actual physical signal with a finite extent (a pulse of light) travels at a different speed. The largest part of the pulse travels at the group velocity v_g, and its earliest part travels at the front velocity v_f.

The phase velocity is important in determining how a light wave travels through a material or from one material to another. It is often represented in terms of a *refractive index*. The refractive index of a material is defined as the ratio of c to the phase velocity v_p in the material: larger indices of refraction indicate lower speeds. The refractive index of a material may depend on the light's frequency, intensity, polarization, or direction of propagation; in many cases, though, it can be treated as a material-dependent constant. The refractive index of air is approximately 1.0003. Denser media, such as water, glass, and diamond, have refractive indexes of around 1.3, 1.5 and 2.4, respectively, for visible light. In exotic materials like Bose–Einstein condensates near absolute zero, the effective speed of light may be only a few metres per second. However, this represents absorption and re-radiation delay between atoms, as do all slower-than-c speeds in material substances. As an extreme example of light "slowing" in matter, two independent teams of physicists claimed to bring light to a "complete standstill" by passing it through a Bose–Einstein condensate of the element rubidium, one team at Harvard University and the Rowland Institute for Science in Cambridge, Mass., and the other at the Harvard–Smithsonian Center for Astrophysics, also in Cambridge. However, the popular description of light being "stopped" in these experiments refers only to light being stored in the excited states of atoms, then re-emitted at an arbitrarily later time, as stimulated by a second laser pulse. During the time it had "stopped," it had ceased to be light. This type of behaviour is generally microscopically true of all transparent media which "slow" the speed of light.

In transparent materials, the refractive index generally is greater than 1, meaning that the phase velocity is less than c. In other materials, it is possible for the refractive index to become smaller than 1 for some frequencies; in some exotic materials it is even possible for the index of refraction to become negative. The requirement that causality is not violated implies that the real and imaginary parts of the dielectric constant of any material, corresponding respectively to the index of refraction and to the attenuation coefficient, are linked by the Kramers–Kronig relations. In practical terms, this means that in a material with refractive index less than 1, the absorption of the wave is so quick that no signal can be sent faster than c.

A pulse with different group and phase velocities (which occurs if the phase velocity is not the same for all the frequencies of the pulse) smears out over time, a process known as dispersion. Certain materials have an exceptionally low (or even zero) group velocity for light waves, a phenomenon called slow light, which has been confirmed in various experiments. The opposite, group velocities exceeding c, has also been shown in experiment. It should even be possible for the group velocity to become infinite or negative, with pulses travelling instantaneously or backwards in time.

None of these options, however, allow information to be transmitted faster than c. It is impossible to transmit information with a light pulse any faster than the speed of the earliest part of the pulse (the front velocity). It can be shown that this is (under certain assumptions) always equal to c.

It is possible for a particle to travel through a medium faster than the phase velocity of light in that medium (but still slower than c). When a charged particle does that in a dielectric material, the electromagnetic equivalent of a shock wave, known as Cherenkov radiation, is emitted.

Practical Effects of Finiteness

The speed of light is of relevance to communications: the one-way and round-trip delay time are greater than zero. This applies from small to astronomical scales. On the other hand, some techniques depend on the finite speed of light, for example in distance measurements.

Small Scales

In supercomputers, the speed of light imposes a limit on how quickly data can be sent between processors. If a processor operates at 1 gigahertz, a signal can only travel a maximum of about 30 centimetres (1 ft) in a single cycle. Processors must therefore be placed close to each other to minimize communication latencies; this can cause difficulty with cooling. If clock frequencies continue to increase, the speed of light will eventually become a limiting factor for the internal design of single chips.

Large Distances on Earth

Given that the equatorial circumference of the Earth is about 40075 km and that c is about 300000 km/s, the theoretical shortest time for a piece of information to travel half the globe along the surface is about 67 milliseconds. When light is travelling around the globe in an optical fibre, the actual transit time is longer, in part because the speed of light is slower by about 35% in an optical fibre, depending on its refractive index n. Furthermore, straight lines rarely occur in global communications situations, and delays are created when the signal passes through an electronic switch or signal regenerator.

Spaceflights and Astronomy

Similarly, communications between the Earth and spacecraft are not instantaneous. There is a brief delay from the source to the receiver, which becomes more noticeable as distances increase. This delay was significant for communications between ground control and Apollo 8 when it became the first manned spacecraft to orbit the Moon: for every question, the ground control station had to wait at least three seconds for the answer to arrive. The communications delay between Earth and Mars can vary between five and twenty minutes depending upon the relative positions of the two planets. As a consequence of this, if a robot on the surface of Mars were to encounter a problem, its human controllers would not be aware of it until at least five minutes later, and possibly up to twenty minutes later; it would then take a further five to twenty minutes for instructions to travel from Earth to Mars.

NASA must wait several hours for information from a probe orbiting Jupiter, and if it needs to correct a navigation error, the fix will not arrive at the spacecraft for an equal amount of time, creating a risk of the correction not arriving in time.

Receiving light and other signals from distant astronomical sources can even take much longer. For example, it has taken 13 billion (13×10^9) years for light to travel to Earth from the faraway galaxies viewed in the Hubble Ultra Deep Field images. Those photographs, taken today, capture images of the galaxies as they appeared 13 billion years ago, when the universe was less than a billion years old. The fact that more distant objects appear to be younger, due to the finite speed of light, allows astronomers to infer the evolution of stars, of galaxies, and of the universe itself.

Astronomical distances are sometimes expressed in light-years, especially in popular science publications and media. A light-year is the distance light travels in one year, around 9461 billion kilometres, 5879 billion miles, or 0.3066 parsecs. In round figures, a light year is nearly 10 trillion kilometres or nearly 6 trillion miles. Proxima Centauri, the closest star to Earth after the Sun, is around 4.2 light-years away.

Distance Measurement

Radar systems measure the distance to a target by the time it takes a radio-wave pulse to return to the radar antenna after being reflected by the target: the distance to the target is half the round-trip transit time multiplied by the speed of light. A Global Positioning System (GPS) receiver measures its distance to GPS satellites based on how long it takes for a radio signal to arrive from each satellite, and from these distances calculates the receiver's position. Because light travels about 300000 kilometres (186000 mi) in one second, these measurements of small fractions of a second must be very precise. The Lunar Laser Ranging Experiment, radar astronomy and the Deep Space Network determine distances to the Moon, planets and spacecraft, respectively, by measuring round-trip transit times.

High-frequency Trading

The speed of light has become important in high-frequency trading, where traders seek to gain minute advantages by delivering their trades to exchanges fractions of a second ahead of other traders. For example, traders have been switching to microwave communications between trading hubs, because of the advantage which microwaves travelling at near to the speed of light in air, have over fibre optic signals which travel 30–40% slower at the speed of light through glass.

Measurement

There are different ways to determine the value of c. One way is to measure the actual speed at which light waves propagate, which can be done in various astronomical and earth-based setups. However, it is also possible to determine c from other physical laws where it appears, for example, by determining the values of the electromagnetic constants ε_0 and μ_0 and using their relation to c. Historically, the most accurate results have been obtained by separately determining the frequency and wavelength of a light beam, with their product equalling c.

In 1983 the metre was defined as "the length of the path travelled by light in vacuum during a time interval of $\frac{1}{299792458}$ of a second", fixing the value of the speed of light at 299792458 m/s by definition, as described below. Consequently, accurate measurements of the speed of light yield an accurate realization of the metre rather than an accurate value of c.

Astronomical Measurements

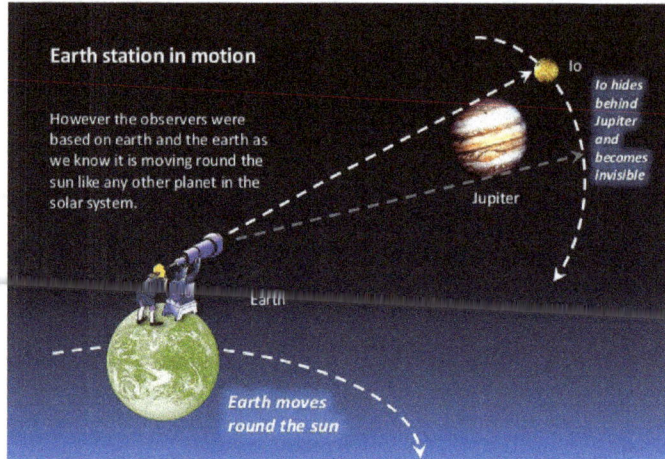

Outer space is a convenient setting for measuring the speed of light because of its large scale and nearly perfect vacuum. Typically, one measures the time needed for light to traverse some reference distance in the solar system, such as the radius of the Earth's orbit. Historically, such measurements could be made fairly accurately, compared to how accurately the length of the reference distance is known in Earth-based units. It is customary to express the results in astronomical units (AU) per day.

Ole Christensen Rømer used an astronomical measurement to make the first quantitative estimate of the speed of light. When measured from Earth, the periods of moons orbiting a distant planet are shorter when the Earth is approaching the planet than when the Earth is receding from it. The distance travelled by light from the planet (or its moon) to Earth is shorter when the Earth is at the point in its orbit that is closest to its planet than when the Earth is at the farthest point in its orbit, the difference in distance being the diameter of the Earth's orbit around the Sun. The observed change in the moon's orbital period is caused by the difference in the time it takes light to traverse the shorter or longer distance. Rømer observed this effect for Jupiter's innermost moon Io and deduced that light takes 22 minutes to cross the diameter of the Earth's orbit.

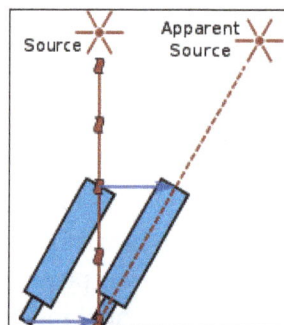

Aberration of light: Light from a distant source appears to be from a different location for a moving telescope due to the finite speed of light.

Another method is to use the aberration of light, discovered and explained by James Bradley in the 18th century. This effect results from the vector addition of the velocity of light arriving from a distant source (such as a star) and the velocity of its observer. A moving observer thus sees the light coming from a slightly different direction and consequently sees the source at a position shifted from its original position. Since the direction of the Earth's velocity changes continuously as the Earth orbits the Sun, this effect causes the apparent position of stars to move around. From the angular difference in the position of stars (maximally 20.5 arcseconds) it is possible to express the speed of light in terms of the Earth's velocity around the Sun, which with the known length of a year can be converted to the time needed to travel from the Sun to the Earth. In 1729, Bradley used this method to derive that light travelled 10,210 times faster than the Earth in its orbit (the modern figure is 10,066 times faster) or, equivalently, that it would take light 8 minutes 12 seconds to travel from the Sun to the Earth.

Astronomical Unit

An astronomical unit (AU) is approximately the average distance between the Earth and Sun. It was redefined in 2012 as exactly 149597870700 m. Previously the AU was not based on the International System of Units but in terms of the gravitational force exerted by the Sun in the framework of classical mechanics. The current definition uses the recommended value in metres for the previous definition of the astronomical unit, which was determined by measurement. This redefinition is analogous to that of the metre, and likewise has the effect of fixing the speed of light to an exact value in astronomical units per second (via the exact speed of light in metres per second).

Previously, the inverse of c expressed in seconds per astronomical unit was measured by comparing the time for radio signals to reach different spacecraft in the Solar System, with their position calculated from the gravitational effects of the Sun and various planets. By combining many such measurements, a best fit value for the light time per unit distance could be obtained. For example, in 2009, the best estimate, as approved by the International Astronomical Union (IAU), was:

light time for unit distance: t_{au} = 499.004783836(10) s

c = 0.00200398880410(4) AU/s = 173.144632674(3) AU/day.

The relative uncertainty in these measurements is 0.02 parts per billion (2×10^{-11}), equivalent to the uncertainty in Earth-based measurements of length by interferometry. Since the metre is defined to be the length travelled by light in a certain time interval, the measurement of the light time in terms of the previous definition of the astronomical unit can also be interpreted as measuring the length of an AU (old definition) in metres.

Time of Flight Techniques

A method of measuring the speed of light is to measure the time needed for light to travel to a mirror at a known distance and back. This is the working principle behind the Fizeau–Foucault apparatus developed by Hippolyte Fizeau and Léon Foucault.

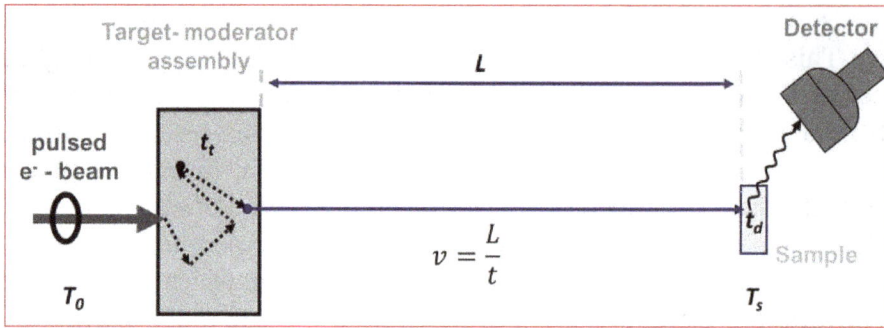

The setup as used by Fizeau consists of a beam of light directed at a mirror 8 kilometres (5 mi) away. On the way from the source to the mirror, the beam passes through a rotating cogwheel. At a certain rate of rotation, the beam passes through one gap on the way out and another on the way back, but at slightly higher or lower rates, the beam strikes a tooth and does not pass through the wheel. Knowing the distance between the wheel and the mirror, the number of teeth on the wheel, and the rate of rotation, the speed of light can be calculated.

The method of Foucault replaces the cogwheel by a rotating mirror. Because the mirror keeps rotating while the light travels to the distant mirror and back, the light is reflected from the rotating mirror at a different angle on its way out than it is on its way back. From this difference in angle, the known speed of rotation and the distance to the distant mirror the speed of light may be calculated.

Nowadays, using oscilloscopes with time resolutions of less than one nanosecond, the speed of light can be directly measured by timing the delay of a light pulse from a laser or an LED reflected from a mirror. This method is less precise (with errors of the order of 1%) than other modern techniques, but it is sometimes used as a laboratory experiment in college physics classes.

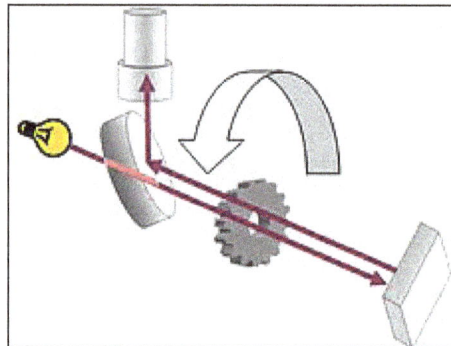

Diagram of the Fizeau apparatus.

Electromagnetic Constants

An option for deriving c that does not directly depend on a measurement of the propagation of electromagnetic waves is to use the relation between c and the vacuum permittivity ε_0 and vacuum permeability μ_0 established by Maxwell's theory: $c^2 = 1/(\varepsilon_0 \mu_0)$. The vacuum permittivity may be determined by measuring the capacitance and dimensions of a capacitor, whereas the value of the vacuum permeability is fixed at exactly $4\pi \times 10^{-7}$ H·m^{-1} through the definition of the ampere. Rosa and Dorsey used this method in 1907 to find a value of 299710±22 km/s.

Cavity Resonance

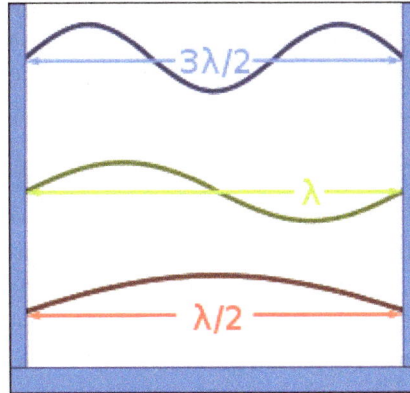

Electromagnetic standing waves in a cavity.

Another way to measure the speed of light is to independently measure the frequency f and wavelength λ of an electromagnetic wave in vacuum. The value of c can then be found by using the relation $c = f\lambda$. One option is to measure the resonance frequency of a cavity resonator. If the dimensions of the resonance cavity are also known, these can be used to determine the wavelength of the wave. In 1946, Louis Essen and A.C. Gordon-Smith established the frequency for a variety of normal modes of microwaves of a microwave cavity of precisely known dimensions. The dimensions were established to an accuracy of about ±0.8 μm using gauges calibrated by interferometry. As the wavelength of the modes was known from the geometry of the cavity and from electromagnetic theory, knowledge of the associated frequencies enabled a calculation of the speed of light.

The Essen–Gordon-Smith result, 299792±9 km/s, was substantially more precise than those found by optical techniques. By 1950, repeated measurements by Essen established a result of 299792.5±3.0 km/s.

A household demonstration of this technique is possible, using a microwave oven and food such as marshmallows or margarine: if the turntable is removed so that the food does not move, it will cook the fastest at the antinodes (the points at which the wave amplitude is the greatest), where it will begin to melt. The distance between two such spots is half the wavelength of the microwaves; by measuring this distance and multiplying the wavelength by the microwave frequency (usually displayed on the back of the oven, typically 2450 MHz), the value of c can be calculated, "often with less than 5% error".

Interferometry

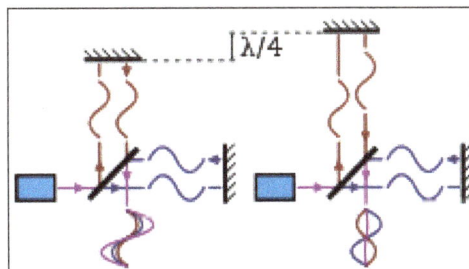

An interferometric determination of length. Left: constructive interference;
Right: destructive interference.

Interferometry is another method to find the wavelength of electromagnetic radiation for determining the speed of light. A coherent beam of light (e.g. from a laser), with a known frequency (f), is split to follow two paths and then recombined. By adjusting the path length while observing the interference pattern and carefully measuring the change in path length, the wavelength of the light (λ) can be determined. The speed of light is then calculated using the equation $c = \lambda f$.

Before the advent of laser technology, coherent radio sources were used for interferometry measurements of the speed of light. However interferometric determination of wavelength becomes less precise with wavelength and the experiments were thus limited in precision by the long wavelength (~0.4 cm (0.16 in)) of the radiowaves. The precision can be improved by using light with a shorter wavelength, but then it becomes difficult to directly measure the frequency of the light. One way around this problem is to start with a low frequency signal of which the frequency can be precisely measured, and from this signal progressively synthesize higher frequency signals whose frequency can then be linked to the original signal. A laser can then be locked to the frequency, and its wavelength can be determined using interferometry. This technique was due to a group at the National Bureau of Standards (NBS) (which later became NIST). They used it in 1972 to measure the speed of light in vacuum with a fractional uncertainty of 3.5×10^{-9}.

First Measurement Attempts

In 1629, Isaac Beeckman proposed an experiment in which a person observes the flash of a cannon reflecting off a mirror about one mile (1.6 km) away. In 1638, Galileo Galilei proposed an experiment, with an apparent claim to having performed it some years earlier, to measure the speed of light by observing the delay between uncovering a lantern and its perception some distance away. He was unable to distinguish whether light travel was instantaneous or not, but concluded that if it were not, it must nevertheless be extraordinarily rapid. In 1667, the Accademia del Cimento of Florence reported that it had performed Galileo's experiment, with the lanterns separated by about one mile, but no delay was observed. The actual delay in this experiment would have been about 11 microseconds.

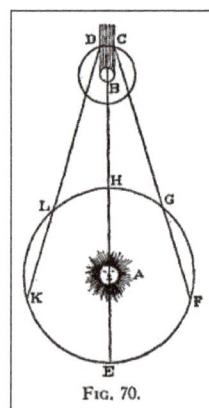

FIG. 70.

Rømer's observations of the occultations of Io from Earth.

The first quantitative estimate of the speed of light was made in 1676 by Rømer. From the observation that the periods of Jupiter's innermost moon Io appeared to be shorter when the Earth was approaching Jupiter than when receding from it, he concluded that light travels at a finite speed, and estimated that it takes light 22 minutes to cross the diameter of Earth's orbit. Christiaan

Huygens combined this estimate with an estimate for the diameter of the Earth's orbit to obtain an estimate of speed of light of 220000 km/s, 26% lower than the actual value.

Isaac Newton reported Rømer's calculations of the finite speed of light and gave a value of "seven or eight minutes" for the time taken for light to travel from the Sun to the Earth (the modern value is 8 minutes 19 seconds). Newton queried whether Rømer's eclipse shadows were coloured; hearing that they were not, he concluded the different colours travelled at the same speed. In 1729, James Bradley discovered stellar aberration. From this effect he determined that light must travel 10,210 times faster than the Earth in its orbit (the modern figure is 10,066 times faster) or, equivalently, that it would take light 8 minutes 12 seconds to travel from the Sun to the Earth.

Connections with Electromagnetism

In the 19th century Hippolyte Fizeau developed a method to determine the speed of light based on time-of-flight measurements on Earth and reported a value of 315000 km/s. His method was improved upon by Léon Foucault who obtained a value of 298000 km/s in 1862. In the year 1856, Wilhelm Eduard Weber and Rudolf Kohlrausch measured the ratio of the electromagnetic and electrostatic units of charge, $1/\sqrt{\varepsilon_o \mu_o}$, by discharging a Leyden jar, and found that its numerical value was very close to the speed of light as measured directly by Fizeau. The following year Gustav Kirchhoff calculated that an electric signal in a resistanceless wire travels along the wire at this speed. In the early 1860s, Maxwell showed that, according to the theory of electromagnetism he was working on, electromagnetic waves propagate in empty space at a speed equal to the above Weber/Kohlrausch ratio, and drawing attention to the numerical proximity of this value to the speed of light as measured by Fizeau, he proposed that light is in fact an electromagnetic wave.

Luminiferous Aether

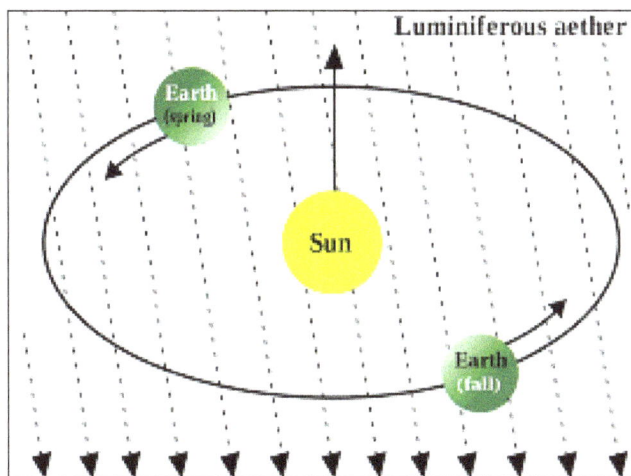

It was thought at the time that empty space was filled with a background medium called the luminiferous aether in which the electromagnetic field existed. Some physicists thought that this aether acted as a preferred frame of reference for the propagation of light and therefore it should be possible to measure the motion of the Earth with respect to this medium, by measuring the isotropy of the speed of light. Beginning in the 1880s several experiments were performed to try to detect this motion, the most famous of which is the experiment performed by Albert A. Michelson and Edward W. Morley in 1887. The detected motion was always less than the observational error. Modern experiments indicate that the two-way speed of light is isotropic (the same in every direction) to within 6 nanometres per second. Because of this experiment Hendrik Lorentz proposed that the motion of the apparatus through the aether may cause the apparatus to contract along its length in the direction of motion, and he further assumed, that the time variable for moving systems must also be changed accordingly ("local time"), which led to the formulation of the Lorentz transformation. Based on Lorentz's aether theory, Henri Poincaré showed that this local time (to first order in v/c) is indicated by clocks moving in the aether, which are synchronized under the assumption of constant light speed. In 1904, he speculated that the speed of light could be a limiting velocity in dynamics, provided that the assumptions of Lorentz's theory are all confirmed. In 1905, Poincaré brought Lorentz's aether theory into full observational agreement with the principle of relativity.

Special Relativity

In 1905 Einstein postulated from the outset that the speed of light in vacuum, measured by a non-accelerating observer, is independent of the motion of the source or observer. Using this and the principle of relativity as a basis he derived the special theory of relativity, in which the speed of light in vacuum c featured as a fundamental constant, also appearing in contexts unrelated to light. This made the concept of the stationary aether (to which Lorentz and Poincaré still adhered) useless and revolutionized the concepts of space and time.

Increased Accuracy of c and Redefinition of the Metre and Second

In the second half of the 20th century much progress was made in increasing the accuracy of measurements of the speed of light, first by cavity resonance techniques and later by laser interferometer techniques. These were aided by new, more precise, definitions of the metre and second. In 1950, Louis Essen determined the speed as 299792.5±1 km/s, using cavity resonance. This value was adopted by the 12th General Assembly of the Radio-Scientific Union in 1957. In 1960, the metre was redefined in terms of the wavelength of a particular spectral line of krypton-86, and, in 1967, the second was redefined in terms of the hyperfine transition frequency of the ground state of caesium-133.

In 1972, using the laser interferometer method and the new definitions, a group at the US National Bureau of Standards in Boulder, Colorado determined the speed of light in vacuum to be $c =$ 299792456.2±1.1 m/s. This was 100 times less uncertain than the previously accepted value. The remaining uncertainty was mainly related to the definition of the metre. As similar experiments found comparable results for c, the 15th General Conference on Weights and Measures in 1975 recommended using the value 299792458 m/s for the speed of light.

Defining the Speed of Light as an Explicit Constant

In 1983 the 17th CGPM found that wavelengths from frequency measurements and a given value for the speed of light are more reproducible than the previous standard. They kept the 1967 definition of second, so the caesium hyperfine frequency would now determine both the second and the metre. To do this, they redefined the metre as: "The metre is the length of the path travelled by light in vacuum during a time interval of 1/299792458 of a second." As a result of this definition, the value of the speed of light in vacuum is exactly 299792458 m/s and has become a defined constant in the SI system of units. Improved experimental techniques that prior to 1983 would have measured the speed of light, no longer affect the known value of the speed of light in SI units, but instead allow a more precise realization of the metre by more accurately measuring the wavelength of Krypton-86 and other light sources.

In 2011, the CGPM stated its intention to redefine all seven SI base units using what it calls "the explicit-constant formulation", where each "unit is defined indirectly by specifying explicitly an exact value for a well-recognized fundamental constant", as was done for the speed of light. It proposed a new, but completely equivalent, wording of the metre's definition: "The metre, symbol m, is the unit of length; its magnitude is set by fixing the numerical value of the speed of light in vacuum to be equal to exactly 299792458 when it is expressed in the SI unit m s^{-1}." This is one of the proposed changes to be incorporated in the next revision of the SI, also termed the *New SI*.

Michelson–Morley Experiment

The Michelson–Morley experiment was an attempt to detect the existence of aether, a supposed medium permeating space that was thought to be the carrier of light waves. The experiment was performed between April and July 1887 by Albert A. Michelson and Edward W. Morley at what is now Case Western Reserve University in Cleveland, Ohio, and published in November of the same year. It compared the speed of light in perpendicular directions, in an attempt to detect the relative motion of matter through the stationary luminiferous aether ("aether wind"). The result was negative, in that Michelson and Morley found no significant difference between the speed of light in the direction of movement through the presumed aether, and the speed at right angles. This result is generally considered to be the first strong evidence against the then-prevalent aether theory, and initiated a line of research that eventually led to special relativity, which rules out a stationary aether. Of this experiment, Einstein wrote, "If the Michelson–Morley experiment had not brought us into serious embarrassment, no one would have regarded the relativity theory as a (halfway) redemption."

Michelson–Morley type experiments have been repeated many times with steadily increasing sensitivity. These include experiments from 1902 to 1905, and a series of experiments in the 1920s. More recent optical resonator experiments confirmed the absence of any aether wind at the 10^{-17} level. Together with the Ives–Stilwell and Kennedy–Thorndike experiments, Michelson–Morley type experiments form one of the fundamental tests of special relativity theory.

Detecting the Aether

Physics theories of the late 19th century assumed that just as surface water waves must have a supporting substance, i.e., a "medium", to move across (in this case water), and audible sound requires a medium to transmit its wave motions (such as air or water), so light must also require a medium, the "luminiferous aether", to transmit its wave motions. Because light can travel through a vacuum, it was assumed that even a vacuum must be filled with aether. Because the speed of light is so great, and because material bodies pass through the *aether* without obvious friction or drag, it was assumed to have a highly unusual combination of properties. Designing experiments to investigate these properties was a high priority of 19th century physics.

Earth orbits around the Sun at a speed of around 30 km/s (18.64 mi/s), or 108,000 km/h (67,000 mph). The Earth is in motion, so two main possibilities were considered: (1) The aether is stationary and only partially dragged by Earth, or (2) the aether is completely dragged by Earth and thus shares its motion at Earth's surface. In addition, James Clerk Maxwell recognized the electromagnetic nature of light and developed what are now called Maxwell's equations, but these equations were still interpreted as describing the motion of waves through an aether, whose state of motion was unknown. Eventually, Fresnel's idea of an (almost) stationary aether was preferred because it appeared to be confirmed by the Fizeau experiment and the aberration of star light.

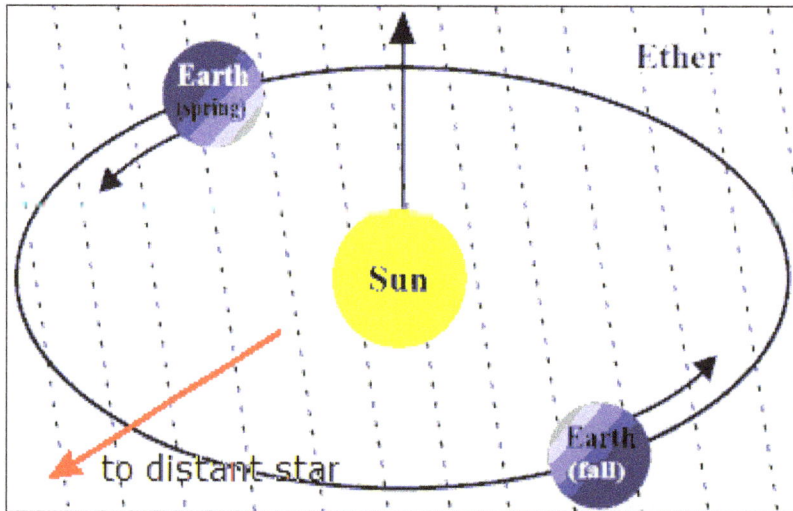

A depiction of the concept of the "aether wind".

According to the stationary and the partially-dragged aether hypotheses, Earth and the aether are in relative motion, implying that a so-called "aether wind" should exist. Although it would be possible, in theory, for the Earth's motion to match that of the aether at one moment in time, it was not possible for the Earth to remain at rest with respect to the aether at all times, because of the variation in both the direction and the speed of the motion. At any given point on the Earth's surface, the magnitude and direction of the wind would vary with time of day and season. By analyzing the return speed of light in different directions at various different times, it was thought to be possible to measure the motion of the Earth relative to the aether. The expected relative difference in the measured speed of light was quite small, given that the velocity of the Earth in its orbit around the Sun has a magnitude of about one hundredth of one percent of the speed of light.

During the mid-19th century, measurements of aether wind effects of first order, i.e., effects proportional to v/c (v being Earth's velocity, c the speed of light) were thought to be possible, but no direct measurement of the speed of light was possible with the accuracy required. For instance, the Fizeau–Foucault apparatus could measure the speed of light to perhaps 5% accuracy, which was quite inadequate for measuring directly a first-order 0.01% change in the speed of light. A number of physicists therefore attempted to make measurements of indirect first-order effects not of the speed of light itself, but of variations in the speed of light. The Hoek experiment, for example, was intended to detect interferometric fringe shifts due to speed differences of oppositely propagating light waves through water at rest. The results of such experiments were all negative. This could be explained by using Fresnel's dragging coefficient, according to which the aether and thus light are partially dragged by moving matter. Partial aether-dragging would thwart attempts to measure any first order change in the speed of light. As pointed out by Maxwell, only experimental arrangements capable of measuring second order effects would have any hope of detecting aether drift, i.e., effects proportional to v^2/c^2. Existing experimental setups, however, were not sensitive enough to measure effects of that size.

Michelson Experiment

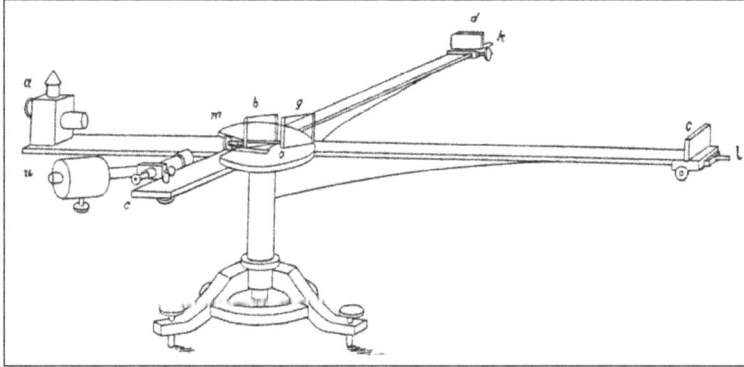

Michelson's 1881 interferometer. Although ultimately it proved incapable of distinguishing between differing theories of aether-dragging, its construction provided important lessons for the design of Michelson and Morley's 1887 instrument.

Michelson had a solution to the problem of how to construct a device sufficiently accurate to detect aether flow. In 1877, while teaching at his alma mater, the United States Naval Academy in Annapolis, Michelson conducted his first known light speed experiments as a part of a classroom demonstration. In 1881, he left active U.S. Naval service while in Germany concluding his studies. In that year, Michelson used a prototype experimental device to make several more measurements.

The device he designed, later known as a Michelson interferometer, sent yellow light from a sodium flame (for alignment), or white light (for the actual observations), through a half-silvered mirror that was used to split it into two beams traveling at right angles to one another. After leaving the splitter, the beams traveled out to the ends of long arms where they were reflected back into the middle by small mirrors. They then recombined on the far side of the splitter in an eyepiece, producing a pattern of constructive and destructive interference whose transverse displacement would depend on the relative time it takes light to transit the longitudinal vs. the transverse arms. If the Earth is traveling through an aether medium, a beam reflecting back and forth parallel to the flow of aether would take longer than a beam reflecting perpendicular to the aether because the time gained from traveling downwind is less than that lost traveling upwind. Michelson expected that the Earth's motion would produce a fringe shift equal to 0.04 fringes—that is, of the separation between areas of the same intensity. He did not observe the expected shift; the greatest average deviation that he measured (in the northwest direction) was only 0.018 fringes; most of his measurements were much less. His conclusion was that Fresnel's hypothesis of a stationary aether with partial aether dragging would have to be rejected, and thus he confirmed Stokes' hypothesis of complete aether dragging.

However, Alfred Potier (and later Hendrik Lorentz) pointed out to Michelson that he had made an error of calculation, and that the expected fringe shift should have been only 0.02 fringes. Michelson's apparatus was subject to experimental errors far too large to say anything conclusive about the aether wind. Definitive measurement of the aether wind would require an experiment with greater accuracy and better controls than the original. Nevertheless, the prototype was successful in demonstrating that the basic method was feasible.

Michelson–Morley Experiment

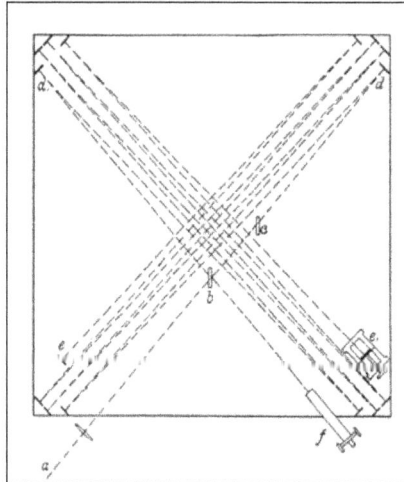

This figure illustrates the folded light path used in the Michelson–Morley interferometer that enabled a path length of 11 m. a is the light source, an oil lamp. b is a beam splitter. c is a compensating plate so that both the reflected and transmitted beams travel through the same amount of glass (important since experiments were run with white light which has an extremely short coherence length requiring precise matching of optical path lengths for fringes to be visible; monochromatic sodium light was used only for initial alignment). d, d' and e are mirrors. e' is a fine adjustment mirror. f is a telescope.

In 1885, Michelson began a collaboration with Edward Morley, spending considerable time and money to confirm with higher accuracy Fizeau's 1851 experiment on Fresnel's drag coefficient, to improve on Michelson's 1881 experiment, and to establish the wavelength of light as a standard of length. At this time Michelson was professor of physics at the Case School of Applied Science, and Morley was professor of chemistry at Western Reserve University (WRU), which shared a campus with the Case School on the eastern edge of Cleveland. Michelson suffered a nervous breakdown in September 1885, from which he recovered by October 1885. Morley ascribed this breakdown to the intense work of Michelson during the preparation of the experiments. In 1886, Michelson and Morley successfully confirmed Fresnel's drag coefficient – this result was also considered as a confirmation of the stationary aether concept.

This result strengthened their hope of finding the aether wind. Michelson and Morley created an improved version of the Michelson experiment with more than enough accuracy to detect this hypothetical effect. The experiment was performed in several periods of concentrated observations between April and July 1887, in the basement of Adelbert Dormitory of WRU.

As shown in figure, the light was repeatedly reflected back and forth along the arms of the interferometer, increasing the path length to 11 m (36 ft). At this length, the drift would be about 0.4 fringes. To make that easily detectable, the apparatus was assembled in a closed room in the basement of the heavy stone dormitory, eliminating most thermal and vibrational effects.

Vibrations were further reduced by building the apparatus on top of a large block of sandstone, about a foot thick and five feet square, which was then floated in a circular trough of mercury. They estimated that effects of about 0.01 fringe would be detectable.

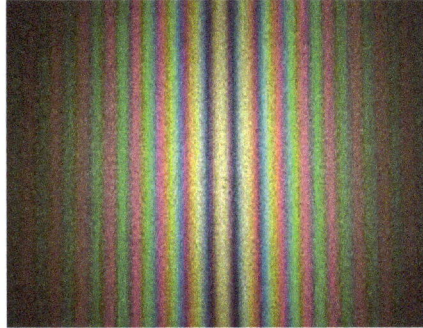

Fringe pattern produced with a Michelson interferometer using white light.
As configured here, the central fringe is white rather than black.

Michelson and Morley and other early experimentalists using interferometric techniques in an attempt to measure the properties of the luminiferous aether, used (partially) monochromatic light only for initially setting up their equipment, always switching to white light for the actual measurements. The reason is that measurements were recorded visually. Purely monochromatic light would result in a uniform fringe pattern. Lacking modern means of environmental temperature control, experimentalists struggled with continual fringe drift even when the interferometer was set up in a basement. Because the fringes would occasionally disappear due to vibrations caused by passing horse traffic, distant thunderstorms and the like, an observer could easily "get lost" when the fringes returned to visibility. The advantages of white light, which produced a distinctive colored fringe pattern, far outweighed the difficulties of aligning the apparatus due to its low coherence length. As Dayton Miller wrote, "White light fringes were chosen for the observations because they consist of a small group of fringes having a central, sharply defined black fringe which forms a permanent zero reference mark for all readings." Use of partially monochromatic light (yellow sodium light) during initial alignment enabled the researchers to locate the position of equal path length, more or less easily, before switching to white light.

The mercury trough allowed the device to turn with close to zero friction, so that once having given the sandstone block a single push it would slowly rotate through the entire range of possible angles to the "aether wind," while measurements were continuously observed by looking through the eyepiece. The hypothesis of aether drift implies that because one of the arms would inevitably turn into the direction of the wind at the same time that another arm was turning perpendicularly to the wind, an effect should be noticeable even over a period of minutes.

The expectation was that the effect would be graphable as a sine wave with two peaks and two troughs per rotation of the device. This result could have been expected because during each full rotation, each arm would be parallel to the wind twice (facing into and away from the wind giving identical readings) and perpendicular to the wind twice. Additionally, due to the Earth's rotation, the wind would be expected to show periodic changes in direction and magnitude during the course of a sidereal day.

Because of the motion of the Earth around the Sun, the measured data were also expected to show annual variations.

Most Famous "Failed" Experiment

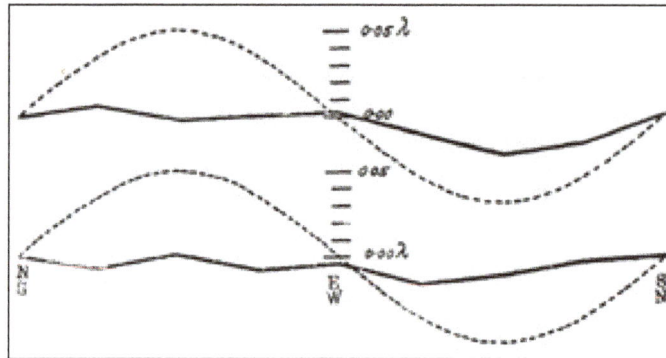

Michelson and Morley's results.

The upper solid line is the curve for their observations at noon, and the lower solid line is that for their evening observations. Note that the theoretical curves and the observed curves are not plotted at the same scale: the dotted curves, in fact, represent only one-eighth of the theoretical displacements.

After all this thought and preparation, the experiment became what has been called the most famous failed experiment in history. Instead of providing insight into the properties of the aether, Michelson and Morley's article in the American Journal of Science reported the measurement to be as small as one-fortieth of the expected displacement, but "since the displacement is proportional to the square of the velocity" they concluded that the measured velocity was "probably less than one-sixth" of the expected velocity of the Earth's motion in orbit and "certainly less than one-fourth." Although this small "velocity" was measured, it was considered far too small to be used as evidence of speed relative to the aether, and it was understood to be within the range of an experimental error that would allow the speed to actually be zero. For instance, Michelson wrote about the "decidedly negative result" in a letter to Lord Rayleigh in August 1887:

> "The Experiments on the relative motion of the earth and ether have been completed and the result decidedly negative. The expected deviation of the interference fringes from the zero should have been 0.40 of a fringe – the maximum displacement was 0.02 and the average much less than 0.01 – and then not in the right place. As displacement is proportional to squares of the relative velocities it follows that if the ether does slip past the relative velocity is less than one sixth of the earth's velocity."

> — Albert Abraham Michelson,

From the standpoint of the then current aether models, the experimental results were conflicting. The Fizeau experiment and its 1886 repetition by Michelson and Morley apparently confirmed the stationary aether with partial aether dragging, and refuted complete aether dragging. On the other hand, the much more precise Michelson–Morley experiment apparently confirmed complete aether dragging and refuted the stationary aether. In addition, the Michelson–Morley null result was further substantiated by the null results of other second-order experiments of different kind, namely the Trouton–Noble experiment and the experiments of Rayleigh and Brace. These problems and their solution led to the development of the Lorentz transformation and special relativity.

After the "failed" experiment Michelson and Morley ceased their aether drift measurements and started to use their newly developed technique to establish the wavelength of light as a standard of length.

Light Path Analysis and Consequences

Observer Resting in the Aether

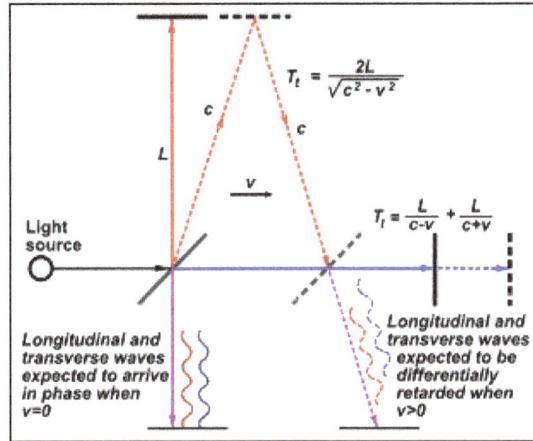

Expected differential phase shift between light traveling the
longitudinal versus the transverse arms of the Michelson–Morley apparatus.

The beam travel time in the longitudinal direction can be derived as follows: Light is sent from the source and propagates with the speed of light c in the aether. It passes through the half-silvered mirror at the origin at T = 0. The reflecting mirror is at that moment at distance L (the length of the interferometer arm) and is moving with velocity v. The beam hits the mirror at time T_1 and thus travels the distance cT_1. At this time, the mirror has traveled the distance vT_1 Thus $cT_1 = L + vT_1$ and consequently the travel time $T_1 = L / (c - v)$. The same consideration applies to the backward journey, with the sign of v reversed, resulting in $cT_2 = L - vT_2$ and $T_2 = L / (c + v)$ The total travel time $T_\ell = T_1 + T_2$ is:

$$T_\ell = \frac{L}{c-v} + \frac{L}{c+v} = \frac{2L}{c} \frac{1}{1 - \frac{v^2}{c^2}} \approx \frac{2L}{c}\left(1 + \frac{v^2}{c^2}\right)$$

Michelson obtained this expression correctly in 1881, however, in transverse direction he obtained the incorrect expression:

$$T_t = \frac{2L}{c},$$

because he overlooked the increased path length in the rest frame of the aether. This was corrected by Alfred Potier and Lorentz. The derivation in the transverse direction can be given as follows (analogous to the derivation of time dilation using a light clock): The beam is propagating at the speed of light c and hits the mirror at time T_3 traveling the distance cT_3. At the same time, the mirror has traveled the distance vT_3 in the x direction. So in order to hit the mirror, the travel path of the beam is L in the y direction (assuming equal-length arms) and vT_3 in the x direction. This inclined travel path follows from the transformation from the interferometer rest frame to the aether rest frame. Therefore, the Pythagorean theorem gives the actual beam travel distance of $\sqrt{L^2 + (vT_3)^2}$. Thus $cT_3 = \sqrt{L^2 + (vT_3)^2}$ and consequently the travel time $T_3 = L / \sqrt{c^2 - v^2}$, which is the same for the backward journey. The total travel time $T_t = 2T_3$ is:

$$T_t = \frac{2L}{\sqrt{c^2 - v^2}} = \frac{2L}{c} \frac{1}{\sqrt{1 - \frac{v^2}{c^2}}} \approx \frac{2L}{c}\left(1 + \frac{v^2}{2c^2}\right)$$

The time difference between T_ℓ and T_t before rotation is given by:

$$T_\ell - T_t = \frac{2}{c}\left(\frac{L}{1 - \frac{v^2}{c^2}} - \frac{L}{\sqrt{1 - \frac{v^2}{c^2}}} \right).$$

By multiplying with c, the corresponding length difference before rotation is:

$$\Delta_1 = 2\left(\frac{L}{1 - \frac{v^2}{c^2}} - \frac{L}{\sqrt{1 - \frac{v^2}{c^2}}} \right),$$

and after rotation:

$$\Delta_2 = 2\left(\frac{L}{\sqrt{1 - \frac{v^2}{c^2}}} - \frac{L}{1 - \frac{v^2}{c^2}} \right).$$

Dividing $\Delta_1 - \Delta_2$ by the wavelength λ, the fringe shift n is found:

$$n = \frac{\Delta_1 - \Delta_2}{\lambda} \approx \frac{2Lv^2}{\lambda c^2}.$$

Since $L \approx 11$ meters and $\lambda \approx 500$ nanometers, the expected fringe shift was n $\approx$ 0.44. So the result would be a delay in one of the light beams that could be detected when the beams were recombined through interference. Any slight change in the spent time would then be observed as a shift in the positions of the interference fringes. The negative result led Michelson to the conclusion that there is no measurable aether drift.

Observer Comoving with the Interferometer

If the same situation is described from the view of an observer co-moving with the interferometer, then the effect of aether wind is similar to the effect experienced by a swimmer, who tries to move with velocity c against a river flowing with velocity v.

In the longitudinal direction the swimmer first moves upstream, so his velocity is diminished due to the river flow to $c - v$. On his way back moving downstream, his velocity is increased to $c + v$. This gives the beam travel times T_1 and T_2.

In the transverse direction, the swimmer has to compensate for the river flow by moving at a certain angle against the flow direction, in order to sustain his exact transverse direction of motion and to reach the other side of the river at the correct location. This diminishes his speed to $\sqrt{c^2 - v^2}$, and gives the beam travel time T_3.

Mirror Reflection

The classical analysis predicted a relative phase shift between the longitudinal and transverse beams which in Michelson and Morley's apparatus should have been readily measurable. What is not often appreciated (since there was no means of measuring it), is that motion through the hypothetical aether should also have caused the two beams to diverge as they emerged from the interferometer by about 10^{-8} radians.

For an apparatus in motion, the classical analysis requires that the beam-splitting mirror be slightly offset from an exact 45° if the longitudinal and transverse beams are to emerge from the apparatus exactly superimposed. In the relativistic analysis, Lorentz-contraction of the beam splitter in the direction of motion causes it to become more perpendicular by precisely the amount necessary to compensate for the angle discrepancy of the two beams.

Length Contraction and Lorentz Transformation

A first step to explaining the Michelson and Morley experiment's null result was found in the FitzGerald–Lorentz contraction hypothesis, now simply called length contraction or Lorentz contraction, first proposed by George FitzGerald and Hendrik Lorentz. According to this law all objects physically contract by L / γ along the line of motion (originally thought to be relative to the aether), $\gamma = 1/\sqrt{1 - v^2 / c^2}$ being the Lorentz factor. This hypothesis was partly motivated by Oliver Heaviside's discovery in 1888 that electrostatic fields are contracting in the line of motion. But since there was no reason at that time to assume that binding forces in matter are of electric origin, length contraction of matter in motion with respect to the aether was considered an Ad hoc hypothesis.

If length contraction of L is inserted into the above formula for T_ℓ, then the light propagation time in the longitudinal direction becomes equal to that in the transverse direction:

$$T_\ell = \frac{2L\sqrt{1 - \dfrac{v^2}{c^2}}}{c} \cdot \frac{1}{1 - \dfrac{v^2}{c^2}} = \frac{2L}{c} \cdot \frac{1}{\sqrt{1 - \dfrac{v^2}{c^2}}} = T_t$$

However, length contraction is only a special case of the more general relation, according to which the transverse length is larger than the longitudinal length by the ratio γ. This can be achieved in many ways. If L_1 is the moving longitudinal length and $L_1' = L_2'$ being the rest lengths, then it is given:

$$\frac{L_2}{L_1} = \frac{L_2'}{\varphi} / \frac{L_1'}{\gamma\varphi} = \gamma.$$

φ can be arbitrarily chosen, so there are infinitely many combinations to explain the Michelson–Morley null result. For instance, if $\varphi = 1$ the relativistic value of length contraction of L_1 occurs, but if $\varphi = 1/\gamma$ then no length contraction but an elongation of L_2 occurs. This hypothesis was later extended by Joseph Larmor, Lorentz and Henri Poincaré, who developed the complete Lorentz transformation including time dilation in order to explain the Trouton–Noble experiment, the Experiments of Rayleigh and Brace, and Kaufmann's experiments. It has the form:

$$x' = \gamma\varphi(x - vt),\; y' = \varphi y,\; z' = \varphi z,\; t' = \gamma\varphi\left(t - \frac{vx}{c^2}\right)$$

It remained to define the value of φ, which was shown by Lorentz to be unity. In general, Poincaré demonstrated that only $\varphi = 1$ allows this transformation to form a group, so it is the only choice compatible with the principle of relativity, i.e., making the stationary aether undetectable. Given this, length contraction and time dilation obtain their exact relativistic values.

Special Relativity

Albert Einstein formulated the theory of special relativity by 1905, deriving the Lorentz transformation and thus length contraction and time dilation from the relativity postulate and the constancy of the speed of light, thus removing the ad hoc character from the contraction hypothesis. Einstein emphasized the kinematic foundation of the theory and the modification of the notion of space and time, with the stationary aether no longer playing any role in his theory. He also pointed out the group character of the transformation. Einstein was motivated by Maxwell's theory of electromagnetism and the lack of evidence for the luminiferous aether.

This allows a more elegant and intuitive explanation of the Michelson–Morley null result. In a comoving frame the null result is self-evident, since the apparatus can be considered as at rest in accordance with the relativity principle, thus the beam travel times are the same. In a frame relative to which the apparatus is moving, the same reasoning applies as described above in "Length contraction and Lorentz transformation", except the word "aether" has to be replaced by "non-comoving inertial frame". Einstein wrote in 1916:

> "Although the estimated difference between these two times is exceedingly small, Michelson and Morley performed an experiment involving interference in which this difference should have been clearly detectable. But the experiment gave a negative result — a fact very perplexing to physicists. Lorentz and FitzGerald rescued the theory from this difficulty by assuming that the motion of the body relative to the æther produces a contraction of the body in the direction of motion, the amount of contraction being just sufficient to compensate for the difference in time mentioned above. The standpoint of the theory of relativity this solution of the difficulty was the right one. But on the basis of the theory of relativity the method of interpretation is incomparably more satisfactory. According to

this theory there is no such thing as a "specially favoured" (unique) co-ordinate system to occasion the introduction of the æther-idea, and hence there can be no æther-drift, nor any experiment with which to demonstrate it. Here the contraction of moving bodies follows from the two fundamental principles of the theory, without the introduction of particular hypotheses; and as the prime factor involved in this contraction we find, not the motion in itself, to which we cannot attach any meaning, but the motion with respect to the body of reference chosen in the particular case in point. Thus for a co-ordinate system moving with the earth the mirror system of Michelson and Morley is not shortened, but it is shortened for a co-ordinate system which is at rest relatively to the sun."

— Albert Einstein

The extent to which the null result of the Michelson–Morley experiment influenced Einstein is disputed. Alluding to some statements of Einstein, many historians argue that it played no significant role in his path to special relativity, while other statements of Einstein probably suggest that he was influenced by it. In any case, the null result of the Michelson–Morley experiment helped the notion of the constancy of the speed of light gain widespread and rapid acceptance.

It was later shown by Howard Percy Robertson and others, that it is possible to derive the Lorentz transformation entirely from the combination of three experiments. First, the Michelson–Morley experiment showed that the speed of light is independent of the orientation of the apparatus, establishing the relationship between longitudinal (β) and transverse (δ) lengths. Then in 1932, Roy Kennedy and Edward Thorndike modified the Michelson–Morley experiment by making the path lengths of the split beam unequal, with one arm being very short. The Kennedy–Thorndike experiment took place for many months as the Earth moved around the sun. Their negative result showed that the speed of light is independent of the velocity of the apparatus in different inertial frames. In addition it established that besides length changes, corresponding time changes must also occur, i.e., it established the relationship between longitudinal lengths (β) and time changes (α). So both experiments do not provide the individual values of these quantities. This uncertainty corresponds to the undefined factor φ as described above. It was clear due to theoretical reasons (the group character of the Lorentz transformation as required by the relativity principle) that the individual values of length contraction and time dilation must assume their exact relativistic form. But a direct measurement of one of these quantities was still desirable to confirm the theoretical results. This was achieved by the Ives–Stilwell experiment, measuring α in accordance with time dilation. Combining this value for α with the Kennedy–Thorndike null result shows that β must assume the value of relativistic length contraction. Combining β with the Michelson–Morley null result shows that δ must be zero. Therefore, the Lorentz transformation with $\varphi = 1$ is an unavoidable consequence of the combination of these three experiments.

Special relativity is generally considered the solution to all negative aether drift (or isotropy of the speed of light) measurements, including the Michelson–Morley null result. Many high precision measurements have been conducted as tests of special relativity and modern searches for Lorentz violation in the photon, electron, nucleon, or neutrino sector, all of them confirming relativity.

Incorrect Alternatives

Michelson initially believed that his experiment would confirm Stokes' theory, according to which the aether was fully dragged in the vicinity of the earth. However, complete aether drag contradicts the observed aberration of light and was contradicted by other experiments as well. In addition, Lorentz showed in 1886 that Stokes's attempt to explain aberration is contradictory.

Furthermore, the assumption that the aether is not carried in the vicinity, but only within matter, was very problematic as shown by the Hammar experiment. Hammar directed one leg of his interferometer through a heavy metal pipe plugged with lead. If aether were dragged by mass, it was theorized that the mass of the sealed metal pipe would have been enough to cause a visible effect. Once again, no effect was seen, so aether-drag theories are considered to be disproven.

Walther Ritz's emission theory (or ballistic theory) was also consistent with the results of the experiment, not requiring aether. The theory postulates that light has always the same velocity in respect to the source. However de Sitter noted that emitter theory predicted several optical effects that were not seen in observations of binary stars in which the light from the two stars could be measured in a spectrometer. If emission theory were correct, the light from the stars should experience unusual fringe shifting due to the velocity of the stars being added to the speed of the light, but no such effect could be seen. It was later shown by J. G. Fox that the original de Sitter experiments were flawed due to extinction, but in 1977 Brecher observed X-rays from binary star systems with similar null results. Also terrestrial tests using particle accelerators have been made that were inconsistent with source dependence of the speed of light. In addition, Emission theory might fail the Ives–Stilwell experiment, but Fox questioned that as well.

Subsequent Experiments

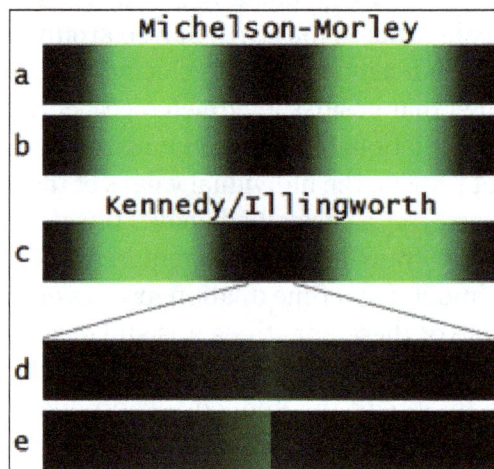

Simulation of the Kennedy/Illingworth refinement of the Michelson–Morley experiment. (a) Michelson–Morley interference pattern in monochromatic mercury light, with a dark fringe precisely centered on the screen. (b) The fringes have been shifted to the left by 1/100 of the fringe spacing. It is extremely difficult to see any difference between this figure and the one above. (c) A small step in one mirror causes two views of the same fringes to be spaced 1/20 of the fringe spacing to the left and to the right of the step. (d) A telescope has been set to view only the central dark band around the mirror step. Note the symmetrical brightening about the center line. (e) The two sets of fringes have been shifted to the left by 1/100 of the fringe spacing. An abrupt discontinuity in luminosity is visible across the step.

Although Michelson and Morley went on to different experiments after their first publication in 1887, both remained active in the field. Other versions of the experiment were carried out with increasing sophistication. Morley was not convinced of his own results, and went on to conduct additional experiments with Dayton Miller from 1902 to 1904. Again, the result was negative within the margins of error.

Miller worked on increasingly larger interferometers, culminating in one with a 32-meter (105 ft) (effective) arm length that he tried at various sites, including on top of a mountain at the Mount Wilson Observatory. To avoid the possibility of the aether wind being blocked by solid walls, his mountaintop observations used a special shed with thin walls, mainly of canvas. From noisy, irregular data, he consistently extracted a small positive signal that varied with each rotation of the device, with the sidereal day, and on a yearly basis. His measurements in the 1920s amounted to approximately 10 km/s (6.2 mi/s) instead of the nearly 30 km/s (18.6 mi/s) expected from the Earth's orbital motion alone. He remained convinced this was due to partial entrainment or aether dragging, though he did not attempt a detailed explanation. He ignored critiques demonstrating the inconsistency of his results and the refutation by the Hammar experiment. Miller's findings were considered important at the time, and were discussed by Michelson, Lorentz and others at a meeting reported in 1928. There was general agreement that more experimentation was needed to check Miller's results. Miller later built a non-magnetic device to eliminate magnetostriction, while Michelson built one of non-expanding Invar to eliminate any remaining thermal effects. Other experimenters from around the world increased accuracy, eliminated possible side effects, or both. So far, no one has been able to replicate Miller's results, and modern experimental accuracies have ruled them out. Roberts has pointed out that the primitive data reduction techniques used by Miller and other early experimenters, including Michelson and Morley, were capable of *creating* apparent periodic signals even when none existed in the actual data. After reanalyzing Miller's original data using modern techniques of quantitative error analysis, Roberts found Miller's apparent signals to be statistically insignificant.

Using a special optical arrangement involving a 1/20 wave step in one mirror, Roy J. Kennedy and K.K. Illingworth converted the task of detecting fringe shifts from the relatively insensitive one of estimating their lateral displacements to the considerably more sensitive task of adjusting the light intensity on both sides of a sharp boundary for equal luminance. If they observed unequal illumination on either side of the step, such as in figure, they would add or remove calibrated weights from the interferometer until both sides of the step were once again evenly illuminated, as in figure. The number of weights added or removed provided a measure of the fringe shift. Different observers could detect changes as little as 1/300 to 1/1500 of a fringe. Kennedy also carried out an experiment at Mount Wilson, finding only about 1/10 the drift measured by Miller and no seasonal effects.

In 1930, Georg Joos conducted an experiment using an automated interferometer with 21-meter-long (69 ft) arms forged from pressed quartz having very low thermal coefficient of expansion, that took continuous photographic strip recordings of the fringes through dozens of revolutions of the apparatus. Displacements of 1/1000 of a fringe could be measured on the photographic plates. No periodic fringe displacements were found, placing an upper limit to the aether wind of 1.5 km/s (0.93 mi/s).

Rapidity

In relativity, rapidity is commonly used as a measure for relativistic velocity. Mathematically, rapidity can be defined as the hyperbolic angle that differentiates two frames of reference in relative motion, each frame being associated with distance and time coordinates.

For one-dimensional motion, rapidities are additive whereas velocities must be combined by Einstein's velocity-addition formula. For low speeds, rapidity and velocity are proportional, but for higher velocities, rapidity takes a larger value, the rapidity of light being infinite.

Using the inverse hyperbolic function artanh, the rapidity w corresponding to velocity v is $w =$ artanh(v / c) where c is the velocity of light. For low speeds, w is approximately v / c. Since in relativity any velocity v is constrained to the interval $-c < v < c$ the ratio v / c satisfies $-1 < v / c < 1$. The inverse hyperbolic tangent has the unit interval $(-1, 1)$ for its domain and the whole real line for its range, and so the interval $-c < v < c$ maps onto $-\infty < w < \infty$.

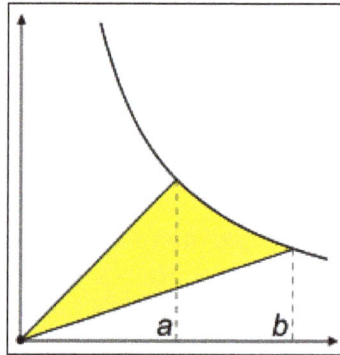

In 1908 Hermann Minkowski explained how the Lorentz transformation could be seen as simply a hyperbolic rotation of the spacetime coordinates, i.e., a rotation through an imaginary angle. This angle therefore represents (in one spatial dimension) a simple additive measure of the velocity between frames. The rapidity parameter replacing velocity was introduced in 1910 by Vladimir Varićak and by E. T. Whittaker. The parameter was named *rapidity* by Alfred Robb and this term was adopted by many subsequent authors, such as Silberstein, Morley and Rindler.

Area of a Hyperbolic Sector

The quadrature of the hyperbola $xy = 1$ by Gregoire de Saint-Vincent established the natural logarithm as the area of a hyperbolic sector, or an equivalent area against an asymptote. In spacetime theory, the connection of events by light divides the universe into Past, Future, or Elsewhere based on a Here and Now. On any line in space, a light beam may be directed left or right. Take the x-axis as the events passed by the right beam and the y-axis as the events of the left beam. Then a resting frame has time along the diagonal $x = y$. The rectangular hyperbola $xy = 1$ can be used to gauge velocities (in the first quadrant). Zero velocity corresponds to (1,1). Any point on the hyperbola has coordinates (e^w, e^{-w}) where w is the rapidity, and is equal to the area of the hyperbolic sector from (1,1) to these coordinates. Many authors refer instead to the unit hyperbola $x^2 - y^2$, using rapidity for parameter, as in the standard spacetime diagram. There the axes are measured by clock and meter-stick, more familiar benchmarks, and the basis of spacetime theory. So the delineation of rapidity as hyperbolic parameter of beam-space is a reference to the seventeenth century origin of our precious transcendental functions, and a supplement to spacetime diagramming.

In One Spatial Dimension

The rapidity w arises in the linear representation of a Lorentz boost as a vector-matrix product:

$$\begin{pmatrix} ct' \\ x' \end{pmatrix} = \begin{pmatrix} \cosh w & -\sinh w \\ -\sinh w & \cosh w \end{pmatrix} \begin{pmatrix} ct \\ x \end{pmatrix} = \Lambda(w) \begin{pmatrix} ct \\ x \end{pmatrix}.$$

The matrix $\Lambda(w)$ is of the type $\begin{pmatrix} p & q \\ q & p \end{pmatrix}$ with p and q satisfying $p^2 - q^2 = 1$, so that (p, q) lies on the unit hyperbola.

Such matrices form the indefinite orthogonal group O(1,1) with one-dimensional Lie algebra spanned by the anti-diagonal unit matrix, showing that the rapidity is the coordinate on this Lie algebra. In matrix exponential notation, $\Lambda(w)$ can be expressed as $\Lambda(w) = e^{Zw}$ where Z is the negative of the anti-diagonal unit matrix:

$$\mathbf{Z} = \begin{pmatrix} 0 & -1 \\ -1 & 0 \end{pmatrix}.$$

It is not hard to prove that:

$$\Lambda(w_1 + w_2) = \Lambda(w_1)\Lambda(w_2).$$

This establishes the useful additive property of rapidity: if A, B and C are frames of reference, then:

$$w_{AC} = w_{AB} + w_{BC}$$

where w_{PQ} denotes the rapidity of a frame of reference Q relative to a frame of reference P. The simplicity of this formula contrasts with the complexity of the corresponding velocity-addition formula.

As we can see from the Lorentz transformation above, the Lorentz factor identifies with cosh w:

$$\gamma = \frac{1}{\sqrt{1 - v^2/c^2}} \equiv \cosh w,$$

so the rapidity w is implicitly used as a hyperbolic angle in the Lorentz transformation expressions using γ and β. We relate rapidities to the velocity-addition formula:

$$u = (u_1 + u_2)/(1 + u_1 u_2 / c^2)$$

by recognizing:

$$\beta_i = \frac{u_i}{c} = \tanh w_i$$

and so:

$$\tanh w = \frac{\tanh w_1 + \tanh w_2}{1 + \tanh w_1 \tanh w_2}$$
$$= \tanh(w_1 + w_2)$$

Proper acceleration (the acceleration 'felt' by the object being accelerated) is the rate of change of rapidity with respect to proper time (time as measured by the object undergoing acceleration itself). Therefore, the rapidity of an object in a given frame can be viewed simply as the velocity of that object as would be calculated non-relativistically by an inertial guidance system on board the object itself if it accelerated from rest in that frame to its given speed.

The product of β and γ appears frequently, and is from the above arguments:

$$\beta\gamma = \sinh w.$$

Exponential and Logarithmic Relations

From the above expressions we have:

$$e^w = \gamma(1 + \beta) = \gamma\left(1 + \frac{v}{c}\right) = \sqrt{\frac{1 + \frac{v}{c}}{1 - \frac{v}{c}}},$$

and thus:

$$e^{-w} = \gamma(1 - \beta) = \gamma\left(1 - \frac{v}{c}\right) = \sqrt{\frac{1 - \frac{v}{c}}{1 + \frac{v}{c}}}.$$

or explicitly:

$$w = \ln\left[\gamma(1 + \beta)\right] = -\ln\left[\gamma(1 - \beta)\right].$$

The Doppler-shift *factor associated with rapidity w is* $k = e^w$.

In more than One Spatial Dimension

The relativistic velocity β is associated to the rapidity w of an object via:

$$\mathfrak{so}(3,1) \supset \operatorname{span}\{K_1, K_2, K_3\} \approx \mathbb{R}^3 \ni \mathbf{w} = \hat{\beta}\tanh^{-1}\beta, \quad \beta \in \mathbb{B}^3,$$

where the vector w is thought of as Cartesian coordinates on the 3-dimensional subspace of the Lie algebra $\mathfrak{o}(3,1) \approx \mathfrak{so}(3,1)$ of the Lorentz group spanned by the boost generators K_1, K_2, K_3 – in complete analogy with the one-dimensional case $\mathfrak{o}(1,1)$ discussed above – and velocity space is represented by the open ball $\mathbb{B}^3$ with radius 1 since $|\beta| < 1$. The latter follows from that c is a limiting velocity in relativity (with units in which c = 1).

The general formula for composition of rapidities is:

$$\mathbf{w} = \hat{\beta}\tanh^{-1}\beta, \quad \beta = \beta_1 \oplus \beta_2,$$

where $\beta_1 \oplus \beta_2$ *refers to* relativistic velocity addition *and* $\hat{\beta}$ *is a unit vector in the direction of* β. This operation is not commutative nor associative. Rapidities $\mathbf{w}_1, \mathbf{w}_2\mathbf{w}_1, \mathbf{w}_2$ *with directions inclined at an angle* θ *have a resultant norm* $w \equiv |\mathbf{w}|$ *(ordinary Euclidean length) given by the* hyperbolic law of cosines,

$$\cosh w = \cosh w_1 \cosh w_2 + \sinh w_1 \sinh w_2 \cos\theta.$$

The geometry on rapidity space is inherited from the hyperbolic geometry on velocity space via the map stated. This geometry, in turn, can be inferred from the addition law of relativistic velocities. Rapidity in two dimensions can thus be usefully visualized using the Poincaré disk. Geodesics correspond to steady accelerations. Rapidity space in three dimensions can in the same way be put in isometry with the hyperboloid model (isometric to the 3-dimensional Poincaré disk (or ball)). This is detailed in geometry of Minkowski space.

The addition of two rapidities results not only in a new rapidity; the resultant total transformation is the composition of the transformation corresponding to the rapidity given above and a rotation parametrized by the vector θ,

$$\Lambda = e^{-i\theta\cdot\mathbf{J}}e^{-i\mathbf{w}\cdot\mathbf{K}},$$

where the physicist convention for the exponential mapping is employed. This is a consequence of the commutation rule:

$$[K_i, K_j] = -i\epsilon_{ijk}J_k,$$

Where $J_k, k = 1, 2, 3,$ are the generators of rotation. This is related to the phenomenon of Thomas precession.

In Experimental Particle Physics

The energy E and scalar momentum |p| of a particle of non-zero (rest) mass m are given by:

$$E = \gamma mc^2$$

$$|\mathbf{p}| = \gamma mv.$$

With the definition of w:

$$w = \operatorname{artanh}\frac{v}{c},$$

and thus with:

$$\cosh w = \cosh\left(\operatorname{artanh}\frac{v}{c}\right) = \frac{1}{\sqrt{1-\dfrac{v^2}{c^2}}} = \gamma$$

$$\sinh w = \sinh\left(\operatorname{artanh}\frac{v}{c}\right) = \frac{\dfrac{v}{c}}{\sqrt{1-\dfrac{v^2}{c^2}}} = \beta\gamma,$$

the energy and scalar momentum can be written as:

$$E = mc^2 \cosh w$$

$$|\mathbf{p}| = mc \sinh w.$$

So rapidity can be calculated from measured energy and momentum by:

$$w = \operatorname{artanh}\frac{|\mathbf{p}|c}{E} = \frac{1}{2}\ln\frac{E+|\mathbf{p}|c}{E-|\mathbf{p}|c}$$

However, experimental particle physicists often use a modified definition of rapidity relative to a beam axis:

$$y = \frac{1}{2}\ln\frac{E+p_z c}{E-p_z c}$$

where p_z is the component of momentum along the beam axis. This is the rapidity of the boost along the beam axis which takes an observer from the lab frame to a frame in which the particle moves only perpendicular to the beam. Related to this is the concept of pseudorapidity.

Maxwell's Equations

Maxwell's equations are a set of coupled partial differential equations that, together with the Lorentz force law, form the foundation of classical electromagnetism, classical optics, and electric circuits. The equations provide a mathematical model for electric, optical, and radio technologies, such as power generation, electric motors, wireless communication, lenses, radar etc. Maxwell's equations describe how electric and magnetic fields are generated by charges, currents, and changes of the fields. An important consequence of the equations is that they demonstrate how fluctuating electric and magnetic fields propagate at a constant speed (c) in a vacuum. Known as electromagnetic radiation, these waves may occur at various wavelengths to produce a spectrum of light from radio waves to *γ-rays*. The equations are named after the physicist and mathematician James Clerk Maxwell, who between 1861 and 1862 published an early form of the equations that included the Lorentz force law. Maxwell first used the equations to propose that light is an electromagnetic phenomenon.

The equations have two major variants. The microscopic Maxwell equations have universal applicability but are unwieldy for common calculations. They relate the electric and magnetic fields to total charge and total current, including the complicated charges and currents in materials at the atomic scale. The "macroscopic" Maxwell equations define two new auxiliary fields that describe the large-scale behaviour of matter without having to consider atomic scale charges and quantum phenomena like spins. However, their use requires experimentally determined parameters for a phenomenological description of the electromagnetic response of materials.

The term "Maxwell's equations" is often also used for equivalent alternative formulations. Versions of Maxwell's equations based on the electric and magnetic potentials are preferred for explicitly solving the equations as a boundary value problem, analytical mechanics, or for use in quantum mechanics. The covariant formulation (on spacetime rather than space and time separately) makes the compatibility of Maxwell's equations with special relativity manifest. Maxwell's equations in curved spacetime, commonly used in high energy and gravitational physics, are compatible with general relativity. In fact, Einstein developed special and general relativity to accommodate the invariant speed of light, a consequence of Maxwell's equations, with the principle that only relative movement has physical consequences.

Since the mid-20th century, it has been understood that Maxwell's equations are not exact, but a classical limit of the fundamental theory of quantum electrodynamics.

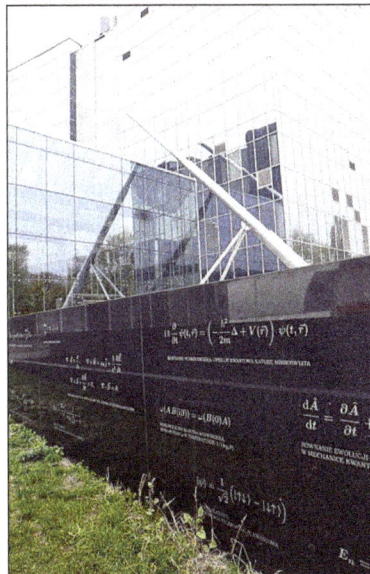

Maxwell's equations as featured on a monument in
front of Warsaw University's Center of New Technologies.

Gauss's Law

Gauss's law describes the relationship between a static electric field and the electric charges that cause it: a static electric field points away from positive charges and towards negative charges, and the net outflow of the electric field through any closed surface is proportional to the charge enclosed by the surface. Picturing the electric field by its field lines, this means the field lines begin at positive electric charges and end at negative electric charges. 'Counting' the number of field lines passing through a closed surface yields the total charge (including bound charge due to polarization of material) enclosed by that surface, divided by dielectricity of free space (the vacuum permittivity).

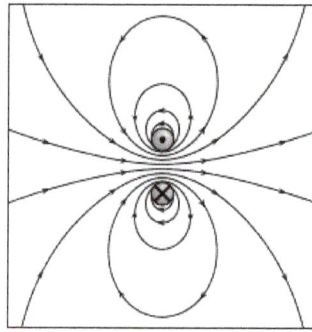

Gauss's law for magnetism: Magnetic field lines never begin nor end but form loops
or extend to infinity as shown here with the magnetic field due to a ring of current.

Gauss's Law for Magnetism

Gauss's law for magnetism states that there are no "magnetic charges" (also called magnetic monopoles), analogous to electric charges. Instead, the magnetic field due to materials is generated by a configuration called a dipole, and the net outflow of the magnetic field through any closed surface is zero. Magnetic dipoles are best represented as loops of current but resemble positive and negative 'magnetic charges', inseparably bound together, having no net 'magnetic charge'. In terms of field lines, this equation states that magnetic field lines neither begin nor end but make loops or extend to infinity and back. In other words, any magnetic field line that enters a given volume must somewhere exit that volume. Equivalent technical statements are that the sum total magnetic flux through any Gaussian surface is zero, or that the magnetic field is a solenoidal vector field.

Faraday's Law

In a geomagnetic storm, a surge in the flux of charged particles temporarily alters Earth's magnetic field, which
induces electric fields in Earth's atmosphere, thus causing surges in electrical power grids.

The Maxwell–Faraday version of Faraday's law of induction describes how a time varying magnetic field creates ("induces") an electric field. In integral form, it states that the work per unit charge required to move a charge around a closed loop equals the rate of decrease of the magnetic flux through the enclosed surface.

The dynamically induced electric field has closed field lines similar to a magnetic field, unless superposed by a static (charge induced) electric field. This aspect of electromagnetic induction is the operating principle behind many electric generators: for example, a rotating bar magnet creates a changing magnetic field, which in turn generates an electric field in a nearby wire.

Ampère's Law with Maxwell's Addition

Magnetic core memory is an application of Ampère's law. Each core stores one bit of data.

Ampère's law with Maxwell's addition states that magnetic fields can be generated in two ways: by electric current (this was the original "Ampère's law") and by changing electric fields (this was "Maxwell's addition", which he called displacement current). In integral form, the magnetic field induced around any closed loop is proportional to the electric current plus displacement current (proportional to the rate of change of electric flux) through the enclosed surface.

Maxwell's addition to Ampère's law is particularly important: it makes the set of equations mathematically consistent for non static fields, without changing the laws of Ampere and Gauss for static fields. However, as a consequence, it predicts that a changing magnetic field induces an electric field and vice versa. Therefore, these equations allow self-sustaining "electromagnetic waves" to travel through empty space.

The speed calculated for electromagnetic waves, which could be predicted from experiments on charges and currents, exactly matches the speed of light; indeed, light *is* one form of electromagnetic radiation (as are X-rays, radio waves, and others). Maxwell understood the connection between electromagnetic waves and light in 1861, thereby unifying the theories of electromagnetism and optics.

Formulation in Terms of Electric and Magnetic Fields

In the electric and magnetic field formulation there are four equations that determine the fields for given charge and current distribution. A separate law of nature, the Lorentz force law, describes how, conversely, the electric and magnetic field act on charged particles and currents. A version of this law was included in the original equations by Maxwell but, by convention, is included no longer. The vector calculus formalism, due to Oliver Heaviside, has become standard. It is manifestly rotation invariant, and therefore mathematically much more transparent than Maxwell's original 20 equations in x,y,z components. The relativistic formulations are even more symmetric and manifestly Lorentz invariant. For the same equations expressed using tensor calculus or differential forms.

The differential and integral equations formulations are mathematically equivalent and are both useful. The integral formulation relates fields within a region of space to fields on the boundary and can often be used to simplify and directly calculate fields from symmetric distributions of charges and currents. On the other hand, the differential equations are purely *local* and are a more natural starting point for calculating the fields in more complicated (less symmetric) situations, for example using finite element analysis.

Key to the Notation

Symbols in bold represent vector quantities, and symbols in *italics* represent scalar quantities, unless otherwise indicated. The equations introduce the electric field, E, a vector field, and the magnetic field, B, a pseudovector field, each generally having a time and location dependence. The sources are:

- The total electric charge density (total charge per unit volume), ρ,

- The total electric current density (total current per unit area), J.

The universal constants appearing in the equations (the first two ones explicitly only in the SI units formulation) are:

- The permittivity of free space, ε_o,

- The permeability of free space, μ_o,

- The speed of light, $C = \dfrac{1}{\sqrt{\varepsilon_0 \mu_0}}$.

Differential Equations

In the differential equations,

- The nabla symbol, ∇, denotes the three-dimensional gradient operator, del,

- The $\nabla \cdot$ symbol denotes the divergence operator,

- The $\nabla \times$ symbol denotes the curl operator.

Integral Equations

In the integral equations,

- Ω is any fixed volume with closed boundary surface $\partial\Omega$,

- Σ is any fixed surface with closed boundary curve $\partial\Sigma$.

Here a fixed volume or surface means that it does not change over time. The equations are correct, complete, and a little easier to interpret with time-independent surfaces. For example, since the surface is time-independent, we can bring the differentiation under the integral sign in Faraday's law:

$$\frac{d}{dt} \iint_{\Sigma} \mathbf{B} \cdot d\mathbf{S} = \iint_{\Sigma} \frac{\partial \mathbf{B}}{\partial t} \cdot d\mathbf{S},$$

Maxwell's equations can be formulated with possibly time-dependent surfaces and volumes by using the differential version and using Gauss and Stokes formula appropriately.

- $\oiint_{\partial\Omega}$ is a surface integral over the boundary surface $\partial\Omega$, with the loop indicating the surface is closed,

- $\iiint_\Omega$ is a volume integral over the volume Ω,

- $\oint_{\partial\Sigma}$ is a line integral around the boundary curve $\partial\Sigma$, with the loop indicating the curve is closed,

- $\iint_\Sigma$ is a surface integral over the surface Σ,

- The total electric charge Q enclosed in Ω is the volume integral over Ω of the charge density ρ:

$$Q = \iiint_\Omega \rho\, dV,$$

where dV is the volume element.

- The net electric current I is the surface integral of the electric current density J passing through a fixed surface, Σ:

$$I = \iint_\Sigma \mathbf{J}\cdot d\mathbf{S},$$

where dS denotes the differential vector element of surface area S, normal to surface Σ. (Vector area is sometimes denoted by A rather than S, but this conflicts with the notation for magnetic potential).

Formulation in SI Units Convention

Name	Integral equations	Differential equations
Gauss's law	$\oiint_{\partial\Omega} \mathbf{E}.d\mathbf{S} = \dfrac{1}{\varepsilon_0}\iiint_\Omega \rho\,dV$	$\nabla\cdot\mathbf{E} = \dfrac{\rho}{\varepsilon_0}$
Gauss's law for magnetism	$\oiint_{\partial\Omega} \mathbf{B}\cdot d\mathbf{S} = 0$	$\nabla\cdot\mathbf{B} = 0$
Maxwell–Faraday equation (Faraday's law of induction)	$\oint_{\partial\Omega} \mathbf{E}\cdot d\mathbf{l} = -\dfrac{d}{dt}\iint_\Sigma \mathbf{B}\cdot d\mathbf{S}$	$\nabla\times\mathbf{E} = -\dfrac{\partial\mathbf{B}}{\partial t}$
Ampère's circuital law (with Maxwell's addition)	$\oint_{\partial\Sigma} \mathbf{B}\cdot d\mathbf{l} = \mu_0\left(\iint_\Sigma \mathbf{J}\cdot d\mathbf{S} + \varepsilon_0\dfrac{d}{dt}\iint_\Sigma \mathbf{E}\cdot d\mathbf{S}\right)$	$\nabla\times\mathbf{B} = \mu_0\left(\mathbf{J} + \varepsilon_0\dfrac{\partial\mathbf{E}}{\partial t}\right)$

Formulation in Gaussian Units Convention

The definitions of charge, electric field, and magnetic field can be altered to simplify theoretical calculation, by absorbing dimensioned factors of ε_0 and μ_0 into the units of calculation, by convention. With a corresponding change in convention for the Lorentz force law this yields the same physics, i.e. trajectories of charged particles, or work done by an electric motor. These definitions are often preferred in theoretical and high energy physics where it is natural to take the electric and magnetic field with the same units, to simplify the appearance of the electromagnetic tensor: the Lorentz covariant object unifying electric and magnetic field would then contain components with uniform unit and dimension. Such modified definitions are conventionally used with the Gaussian (CGS) units. Using these definitions and conventions, colloquially "in Gaussian units", the Maxwell equations become:

Name	Integral equations	Differential equations
Gauss's law	$\oiint_{\partial\Omega} \mathbf{E}{\cdot}d\mathbf{S} = 4\pi \iiint_{\Omega} \rho \, dV$	$\nabla{\cdot}\mathbf{E} = 4\pi\rho$
Gauss's law for magnetism	$\oiint_{\partial\Omega} \mathbf{B}{\cdot}d\mathbf{S} = 0$	$\nabla{\cdot}\mathbf{B} = 0$
Maxwell–Faraday equation (Faraday's law of induction)	$\oint_{\partial\Sigma} \mathbf{E}{\cdot}d\ell = -\dfrac{1}{c}\dfrac{d}{dt}\iint_{\Sigma} \mathbf{B}{\cdot}d\mathbf{S}$	$\nabla\times\mathbf{E} = -\dfrac{1}{c}\dfrac{\partial\mathbf{B}}{\partial t}$
Ampère's circuital law (with Maxwell's addition)	$\oint_{\partial\Sigma} \mathbf{B}{\cdot}d\ell = \dfrac{1}{c}\left(4\pi\iint_{\Sigma} \mathbf{J}{\cdot}d\mathbf{S} + \dfrac{d}{dt}\iint_{\Sigma} \mathbf{E}{\cdot}d\mathbf{S} \right)$	$\nabla\times\mathbf{B} = \dfrac{1}{c}\left(4\pi\mathbf{J} + \dfrac{\partial\mathbf{E}}{\partial t} \right)$

The equations are particularly readable when length and time are measured in compatible units like seconds and lightseconds i.e. in units such that c = 1 unit of length/unit of time. Ever since 1983, metres and seconds are compatible except for historical legacy since *by definition c = 299 792 458 m/s (≈ 1.0 feet/nanosecond)*.

Further cosmetic changes, called rationalisations, are possible by absorbing factors of 4π depending on whether we want Coulomb's law or Gauss's law to come out nicely. In theoretical physics it is often useful to choose units such that Planck's constant, the elementary charge, and even Newton's constant are 1.

Relationship between Differential and Integral Formulations

The equivalence of the differential and integral formulations are a consequence of the Gauss divergence theorem *and the* Kelvin–Stokes theorem.

Flux and Divergence

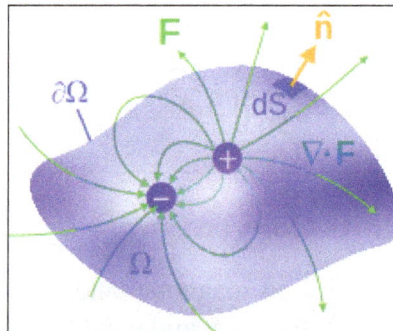

Volume Ω and its closed boundary ∂Ω, containing (respectively enclosing) a source (+) and sink (−) of a vector field F. Here, F could be the E field with source electric charges, but notthe B field, which has no magnetic charges as shown. The outward unit normal is n.

According to the (purely mathematical) Gauss divergence theorem, the electric flux *through the boundary surface $\partial\Omega$ can be rewritten as:*

$$\oiint_{\partial\Sigma} \mathbf{E}\cdot d\mathbf{S} = \iiint_{\Omega} \nabla\cdot\mathbf{E}\, dV$$

The integral version of Gauss's equation can thus be rewritten as:

$$\iiint_{\Omega}\left(\nabla\cdot\mathbf{E} - \frac{\rho}{\epsilon_0}\right) dV = 0$$

Since Ω is arbitrary (e.g. an arbitrary small ball with arbitrary center), this is satisfied iff *the integrand is zero*. This is the differential equations formulation of Gauss equation up to a trivial rearrangement.

Similarly rewriting the magnetic flux in Gauss's law for magnetism in integral form gives:

$$\oiint_{\partial\Omega} \mathbf{B}\cdot d\mathbf{S} = \iiint_{\Omega} \nabla\cdot\mathbf{B}\, dV = 0.$$

which is satisfied for all Ω iff $\nabla\cdot\mathbf{B} = 0$.

Circulation and Curl

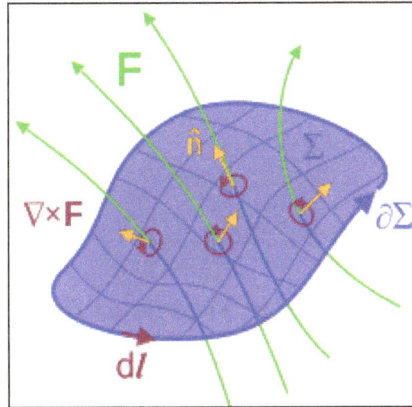

Surface Σ with closed boundary $\partial\Sigma$. **F** *could be the* **E** *or* **B** *fields. Again,* **n** is the unit normal.
(The curl of a vector field doesn't literally look like the "circulations", this is a heuristic depiction.)

By the Kelvin–Stokes theorem we can rewrite the line integrals of the fields around the closed boundary curve $\partial\Sigma$ to an integral of the "circulation of the fields" (i.e. their curls) over a surface it bounds, i.e:

$$\oint_{\partial\Sigma} \mathbf{B}\cdot d\ell = \iint_{\Sigma} (\nabla\times\mathbf{B})\cdot d\mathbf{S},$$

Hence the modified Ampere law in integral form can be rewritten as:

$$\iint_{\Sigma}\left(\nabla\times\mathbf{B} - \mu_0\left(\mathbf{J} + \epsilon_0\frac{\partial\mathbf{E}}{\partial t}\right)\right)\cdot d\mathbf{S} = 0.$$

Since Σ can be chosen arbitrarily, e.g. as an arbitrary small, arbitrary oriented, and arbitrary centered disk, we conclude that the integrand is zero iff Ampere's modified law in differential equations form is satisfied. The equivalence of Faraday's law in differential and integral form follows likewise.

The line integrals and curls are analogous to quantities in classical fluid dynamics: the circulation of a fluid is the line integral of the fluid's flow velocity field around a closed loop, and the vorticity of the fluid is the curl of the velocity field.

Charge Conservation

The invariance of charge can be derived as a corollary of Maxwell's equations. The left-hand side of the modified Ampere's Law has zero divergence by the div–curl identity. Expanding the divergence of the right-hand side, interchanging derivatives, and applying Gauss's law gives:

$$0 = \nabla \cdot \nabla \times \mathbf{B} = \mu_0 \left(\nabla \cdot \mathbf{J} + \varepsilon_0 \frac{\partial}{\partial t} \nabla \cdot \mathbf{E} \right) = \mu_0 \left(\nabla \cdot \mathbf{J} + \frac{\partial \rho}{\partial t} \right)$$

i.e:

$$\frac{\partial \rho}{\partial t} + \nabla \cdot \mathbf{J} = 0.$$

By the Gauss Divergence Theorem, this means the rate of change of charge in a fixed volume equals the net current flowing through the boundary:

$$\frac{d}{dt} Q_\Omega = \frac{d}{dt} \iiint_\Omega \rho \, dV = -\iint_{\partial\Omega} \mathbf{J} \cdot d\mathbf{S} = -I_{\partial\Omega}.$$

In particular, in an isolated system the total charge is conserved.

Vacuum Equations, Electromagnetic Waves and Speed of Light

In a region with no charges ($\rho = 0$) *and no currents* ($\mathbf{J} = 0$), such as in a vacuum, Maxwell's equations reduce to:

$$\nabla \cdot \mathbf{E} = 0 \quad \nabla \times \mathbf{E} = -\frac{\partial \mathbf{B}}{\partial t},$$

$$\nabla \cdot \mathbf{B} = 0 \quad \nabla \times \mathbf{B} = \mu_0 \varepsilon_0 \frac{\partial \mathbf{E}}{\partial t}.$$

Taking the curl ($\nabla \times$) of the curl equations, and using the curl of the curl identity we obtain:

$$\mu_0 \varepsilon_0 \frac{\partial^2 \mathbf{E}}{\partial t^2} - \nabla^2 \mathbf{E} = 0$$

$$\mu_0 \varepsilon_0 \frac{\partial^2 \mathbf{B}}{\partial t^2} - \nabla^2 \mathbf{B} = 0$$

The quantity $\mu_0\varepsilon_0$ has the dimension of (time/length)². Defining $c = (\mu_0\varepsilon_0)^{-1/2}$, the equations above have the form of the standard wave equations:

$$\frac{1}{c^2}\frac{\partial^2 \mathbf{E}}{\partial t^2} - \nabla^2\mathbf{E} = 0$$

$$\frac{1}{c^2}\frac{\partial^2 \mathbf{B}}{\partial t^2} - \nabla^2\mathbf{B} = 0$$

Already during Maxwell's lifetime, it was found that the known values for ε_0 and μ_0 give $c \approx 2.998\times10^8$ m/s, then already known to be the speed of light in free space. This led him to propose that light and radio waves were propagating electromagnetic waves, since amply confirmed. In the old SI system of units, the values of $\mu_0 = 4\pi\times10^{-7}$ and $c = 299792458$ m/s are defined constants, (which means that by definition $\varepsilon_0 = 8.854...\times10^{-12}$ F/m)) that define the ampere and the metre. In the new SI system, only c keeps its defined value, and the electron charge gets a defined value.

In materials with relative permittivity, ε_r, and relative permeability, μ_r, the phase velocity of light becomes:

$$v_p = \frac{1}{\sqrt{\mu_0\mu_r\varepsilon_0\varepsilon_r}}$$

which is usually less than c.

In addition, E and B are perpendicular to each other and to the direction of wave propagation, and are in phase with each other. A sinusoidal plane wave is one special solution of these equations. Maxwell's equations explain how these waves can physically propagate through space. The changing magnetic field creates a changing electric field through Faraday's law. In turn, that electric field creates a changing magnetic field through Maxwell's addition to Ampère's law. This perpetual cycle allows these waves, now known as electromagnetic radiation, to move through space at velocity c.

Macroscopic Formulation

The above equations are the "microscopic" version of Maxwell's equations, expressing the electric and the magnetic fields in terms of the (possibly atomic-level) charges and currents present. This is sometimes called the "general" form, but the macroscopic version below is equally general, the difference being one of bookkeeping.

The microscopic version is sometimes called "Maxwell's equations in a vacuum": this refers to the fact that the material medium is not built into the structure of the equations, but appears only in the charge and current terms. The microscopic version was introduced by Lorentz, who tried to use it to derive the macroscopic properties of bulk matter from its microscopic constituents.

"Maxwell's macroscopic equations", also known as Maxwell's equations in matter, are more similar to those that Maxwell introduced himself.

Name	Integral equations (SI convention)	Differential equations (SI convention)	Differential equations (Gaussian convention)
Gauss's law	$\oiint_{\partial\Omega} \mathbf{D}\cdot d\mathbf{S} = \iiint_{\Omega} \rho_f\, dV$	$\nabla\cdot\mathbf{D} = \rho_f$	$\nabla\cdot\mathbf{D} = 4\pi\rho_f$
Gauss's law for magnetism	$\oiint_{\partial\Omega} \mathbf{B}\cdot d\mathbf{S} = 0$	$\nabla\cdot\mathbf{B} = 0$	$\nabla\cdot\mathbf{B} = 0$
Maxwell–Faraday equation (Faraday's law of induction)	$\oint_{\partial\Sigma} \mathbf{E}\cdot d\boldsymbol{\ell} = -\dfrac{d}{dt}\iint_{\Sigma} \mathbf{B}\cdot d\mathbf{S}$	$\nabla\times\mathbf{E} = -\dfrac{\partial\mathbf{B}}{\partial t}$	$\nabla\times\mathbf{E} = -\dfrac{1}{c}\dfrac{\partial\mathbf{B}}{\partial t}$
Ampère's circuital law (with Maxwell's addition)	$\oint_{\partial\Sigma} \mathbf{H}\cdot d\boldsymbol{\ell} = \iint_{\Sigma} \mathbf{J}_f\cdot d\mathbf{S} + \dfrac{d}{dt}\iint_{\Sigma} \mathbf{D}\cdot d\mathbf{S}$	$\nabla\times\mathbf{H} = \mathbf{J}_f + \dfrac{\partial\mathbf{D}}{\partial t}$	$\nabla\times\mathbf{H} = \dfrac{1}{c}\left(4\pi\mathbf{J}_f + \dfrac{\partial\mathbf{D}}{\partial t}\right)$

In the "macroscopic" equations, the influence of bound charge Q_b and bound current I_b is incorporated into the displacement field D and the magnetizing field H, while the equations depend only on the free charges Q_f and free currents I_f. This reflects a splitting of the total electric charge Q and current I (and their densities ρ and J) into free and bound parts:

$$Q = Q_f + Q_b = \iiint_{\Omega}(\rho_f + \rho_b)\,dV = \iiint_{\Omega}\rho\,dV$$

$$I = I_f + I_b = \iint_{\Sigma}(\mathbf{J}_f + \mathbf{J}_b)\cdot d\mathbf{S} = \iint_{\Sigma}\mathbf{J}\cdot d\mathbf{S}$$

The cost of this splitting is that the additional fields D and H need to be determined through phenomenological constituent equations relating these fields to the electric field E and the magnetic field B, together with the bound charge and current.

Bound Charge and Current

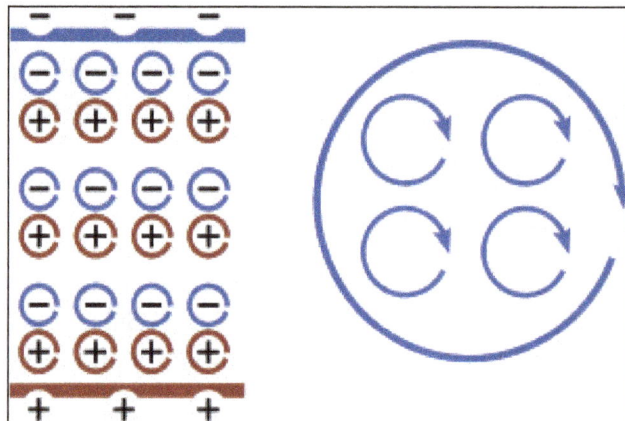

Left: A schematic view of how an assembly of microscopic dipoles produces opposite surface charges as shown at top and bottom. Right: How an assembly of microscopic current loops add together to produce a macroscopically circulating current loop.

When an electric field is applied to a dielectric material its molecules respond by forming microscopic electric dipoles – their atomic nuclei move a tiny distance in the direction of the field, while their electrons move a tiny distance in the opposite direction. This produces a macroscopic bound charge in the material even though all of the charges involved are bound to individual molecules.

For example, if every molecule responds the same, similar to that shown in the figure, these tiny movements of charge combine to produce a layer of positive bound charge on one side of the material and a layer of negative charge on the other side. The bound charge is most conveniently described in terms of the polarization P of the material, its dipole moment per unit volume. If P is uniform, a macroscopic separation of charge is produced only at the surfaces where P enters and leaves the material. For non-uniform P, a charge is also produced in the bulk.

Somewhat similarly, in all materials the constituent atoms exhibit magnetic moments that are intrinsically linked to the angular momentum of the components of the atoms, most notably their electrons. The connection to angular momentum suggests the picture of an assembly of microscopic current loops. Outside the material, an assembly of such microscopic current loops is not different from a macroscopic current circulating around the material's surface, despite the fact that no individual charge is traveling a large distance. These bound currents can be described using the magnetization M.

The very complicated and granular bound charges and bound currents, therefore, can be represented on the macroscopic scale in terms of P and M, which average these charges and currents on a sufficiently large scale so as not to see the granularity of individual atoms, but also sufficiently small that they vary with location in the material. As such, Maxwell's macroscopic equations ignore many details on a fine scale that can be unimportant to understanding matters on a gross scale by calculating fields that are averaged over some suitable volume.

Auxiliary Fields, Polarization and Magnetization

The definitions (not constitutive relations) of the auxiliary fields are:

$$\mathbf{D}(\mathbf{r},t) = \varepsilon_0 \mathbf{E}(\mathbf{r},t) + \mathbf{P}(\mathbf{r},t)$$

$$\mathbf{H}(\mathbf{r},t) = \frac{1}{\mu_0}\mathbf{B}(\mathbf{r},t) - \mathbf{M}(\mathbf{r},t)$$

where P is the polarization field and M is the magnetization field, which are defined in terms of microscopic bound charges and bound currents respectively. The macroscopic bound charge density ρ_b and bound current density $\mathbf{J}_b$ in terms of polarization P and magnetization M are then defined as:

$$\rho_b = -\nabla \cdot \mathbf{P}$$

$$\mathbf{J}_b = \nabla \times \mathbf{M} + \frac{\partial \mathbf{P}}{\partial t}$$

If we define the total, bound, and free charge and current density by:

$$\rho = \rho_b + \rho_f,$$

$$\mathbf{J} = \mathbf{J}_b + \mathbf{J}_f,$$

and use the defining relations above to eliminate D, and H, the "macroscopic" Maxwell's equations reproduce the "microscopic" equations.

Constitutive Relations

In order to apply 'Maxwell's macroscopic equations', it is necessary to specify the relations between displacement field D and the electric field E, as well as the magnetizing field H and the magnetic field B. Equivalently, we have to specify the dependence of the polarization P (hence the bound charge) and the magnetization M (hence the bound current) on the applied electric and magnetic field. The equations specifying this response are called constitutive relations. For real-world materials, the constitutive relations are rarely simple, except approximately, and usually determined by experiment.

For materials without polarization and magnetization, the constitutive relations are:

$$\mathbf{D} = \varepsilon_0 \mathbf{E}, \quad \mathbf{H} = \frac{1}{\mu_0} \mathbf{B}$$

where ε_0 is the permittivity of free space and μ_o the permeability of free space. Since there is no bound charge, the total and the free charge and current are equal.

An alternative viewpoint on the microscopic equations is that they are the macroscopic equations together with the statement that vacuum behaves like a perfect linear "material" without additional polarization and magnetization. More generally, for linear materials the constitutive relations are:

$$\mathbf{D} = \varepsilon \mathbf{E}, \quad \mathbf{H} = \frac{1}{\mu} \mathbf{B}$$

where ε is the permittivity and μ the permeability of the material. For the displacement field D the linear approximation is usually excellent because for all but the most extreme electric fields or temperatures obtainable in the laboratory (high power pulsed lasers) the interatomic electric fields of materials of the order of 10^{11} V/m are much higher than the external field. For the magnetizing field H, however, the linear approximation can break down in common materials like iron leading to phenomena like hysteresis. Even the linear case can have various complications, however.

- For homogeneous materials, ε and μ are constant throughout the material, while for inhomogeneous materials they depend on location within the material (and perhaps time).

- For isotropic materials, ε and μ are scalars, while for anisotropic materials (e.g. due to crystal structure) they are tensors.

- Materials are generally dispersive, so ε and μ depend on the frequency of any incident EM waves.

Even more generally, in the case of non-linear materials, D and P are not necessarily proportional to E, similarly H or M is not necessarily proportional to B. In general D and Hdepend on both E and B, on location and time, and possibly other physical quantities.

In applications one also has to describe how the free currents and charge density behave in terms of E and B possibly coupled to other physical quantities like pressure, and the mass, number density, and velocity of charge-carrying particles. E.g., the original equations given by Maxwell included Ohm's law in the form:

$$\mathbf{J}_f = \sigma \mathbf{E}.$$

Alternative Formulations

Following is a summary of some of the numerous other mathematical formalisms to write the microscopic Maxwell's equations, with the columns separating the two homogeneous Maxwell equations from the two inhomogeneous ones involving charge and current. Each formulation has versions directly in terms of the electric and magnetic fields, and indirectly in terms of the electrical potential φ and the vector potential A. Potentials were introduced as a convenient way to solve the homogeneous equations, but it was thought that all observable physics was contained in the electric and magnetic fields (or relativistically, the Faraday tensor). The potentials play a central role in quantum mechanics, however, and act quantum mechanically with observable consequences even when the electric and magnetic fields vanish (Aharonov–Bohm effect).

Each table describes one formalism. SI units are used throughout.

Vector Calculus

Formulation	Homogeneous equations	Inhomogeneous equations
Fields 3D Euclidean space + time	$\nabla \cdot \mathbf{B} = 0$ $\nabla \times \mathbf{E} + \dfrac{\partial \mathbf{B}}{\partial t} = 0$	$\nabla \cdot \mathbf{E} = \dfrac{\rho}{\varepsilon_0}$ $\nabla \times \mathbf{B} - \dfrac{1}{c^2}\dfrac{\partial \mathbf{E}}{\partial t} = \mu_0 \mathbf{J}$
Potentials (any gauge) 3D Euclidean space + time	$\mathbf{B} = \nabla \times \mathbf{A}$ $\mathbf{E} = -\nabla \varphi - \dfrac{\partial \mathbf{A}}{\partial t}$	$-\nabla^2 \varphi - \dfrac{\partial}{\partial t}(\nabla \cdot \mathbf{A}) = \dfrac{\rho}{\varepsilon_0}$ $\left(-\nabla^2 + \dfrac{1}{c^2}\dfrac{\partial^2}{\partial t^2}\right)\mathbf{A} + \nabla\left(\nabla \cdot \mathbf{A} + \dfrac{1}{c^2}\dfrac{\partial \varphi}{\partial t}\right) = \mu_0 \mathbf{J}$
Potentials (Lorenz gauge) 3D Euclidean space + time	$\mathbf{B} = \nabla \times \mathbf{A}$ $\mathbf{E} = -\nabla \varphi - \dfrac{\partial \mathbf{A}}{\partial t}$ $\nabla \cdot \mathbf{A} = -\dfrac{1}{c^2}\dfrac{\partial \varphi}{\partial t}$	$\left(-\nabla^2 + \dfrac{1}{c^2}\dfrac{\partial^2}{\partial t^2}\right)\varphi = \dfrac{\rho}{\varepsilon_0}$ $\left(-\nabla^2 + \dfrac{1}{c^2}\dfrac{\partial^2}{\partial t^2}\right)\mathbf{A} = \mu_0 \mathbf{J}$

Tensor Calculus

Formulation	Homogeneous equations	Inhomogeneous equations
Fields space + time spatial metric independent of time	$\partial_{[i}B_{jk]} =$ $\nabla_{[i}B_{jk]} = 0$ $\partial_{[i}E_{j]} + \dfrac{\partial B_{ij}}{\partial t} =$ $\nabla_{[i}E_{j]} + \dfrac{\partial B_{ij}}{\partial t} = 0$	$\dfrac{1}{\sqrt{h}}\partial_i\sqrt{h}E^i =$ $\nabla_i E^i = \dfrac{\rho}{\epsilon_0}$ $-\dfrac{1}{\sqrt{h}}\partial_i\sqrt{h}B^{ij} - \dfrac{1}{c^2}\dfrac{\partial}{\partial t}E^j =$ $-\nabla_i B^{ij} - \dfrac{1}{c^2}\dfrac{\partial E^j}{\partial t} = \mu_0 J^j$
Potentials space (with topological restrictions) + time spatial metric independent of time	$B_{ij} = \partial_{[i}A_{j]}$ $\quad = \nabla_{[i}A_{j]}$ $E_i = -\dfrac{\partial A_i}{\partial t} - \partial_i\phi$ $\quad = -\dfrac{\partial A_i}{\partial t} - \nabla_i\phi$	$-\dfrac{1}{\sqrt{h}}\partial_i\sqrt{h}\left(\partial^i\phi + \dfrac{\partial A^i}{\partial t}\right) =$ $-\nabla_i\nabla^i\phi - \dfrac{\partial}{\partial t}\nabla_i A^i = \dfrac{\rho}{\epsilon_0}$ $-\dfrac{1}{\sqrt{h}}\partial_i\sqrt{h}h^{im}h^{jn}\partial_{[m}A_{n]} + \dfrac{1}{c^2}\dfrac{\partial}{\partial t}\left(\dfrac{\partial A^j}{\partial t} + \partial^j\phi\right) =$ $-\nabla_i\nabla^i A^j + \dfrac{1}{c^2}\dfrac{\partial^2 A^j}{\partial t^2} + R_i^j A^i + \nabla^j\left(\nabla_i A^i + \dfrac{1}{c^2}\dfrac{\partial\phi}{\partial t}\right) = \mu_0 J^j$
Potentials (Lorenz gauge) space (with topological restrictions) + time spatial metric independent of time	$B_{ij} = \partial_{[i}A_{j]}$ $\quad = \nabla_{[i}A_{j]}$ $E_i = -\dfrac{\partial A_i}{\partial t} - \partial_i\phi$ $\quad = -\dfrac{\partial A_i}{\partial t} - \nabla_i\phi$ $\nabla_i A^i = -\dfrac{1}{c^2}\dfrac{\partial\phi}{\partial t}$	$-\nabla_i\nabla^i\phi + \dfrac{1}{c^2}\dfrac{\partial^2\phi}{\partial t^2} = \dfrac{\rho}{\epsilon_0}$ $-\nabla_i\nabla^i A^j + \dfrac{1}{c^2}\dfrac{\partial^2 A^j}{\partial t^2} + R_i^j A^i = \mu_0 J^j$

Differential Forms

Formulation	Homogeneous equations	Inhomogeneous equations
Fields Any space + time	$dB = 0$ $dE + \dfrac{\partial B}{\partial t} = 0$	$d*E = \dfrac{\rho}{\epsilon_0}$ $d*B - \dfrac{1}{c^2}\dfrac{\partial *E}{\partial t} = \mu_0 J$
Potentials (any gauge) Any space (with topological restrictions) + time	$B = dA$ $E = -d\phi - \dfrac{\partial A}{\partial t}$	$-d*\left(d\phi + \dfrac{\partial A}{\partial t}\right) = \dfrac{\rho}{\epsilon_0}$ $d*dA + \dfrac{1}{c^2}\dfrac{\partial}{\partial t}*\left(d\phi + \dfrac{\partial A}{\partial t}\right) = \mu_0 J$

Potential (Lorenz Gauge)	$B = dA$	$*\left(-\Delta\phi + \dfrac{1}{c^2}\dfrac{\partial^2}{\partial t^2}\phi\right) = \dfrac{\rho}{\epsilon_0}$
Any space (with topological restrictions) + time	$E = -d\phi - \dfrac{\partial A}{\partial t}$	
spatial metric independent of time	$d * A = -*\dfrac{1}{c^2}\dfrac{\partial\phi}{\partial t}$	$*\left(-\Delta A + \dfrac{1}{c^2}\dfrac{\partial^2 A}{\partial^2 t}\right) = \mu_0 J$

Relativistic Formulations

The Maxwell equations can also be formulated on a spacetime-like Minkowski space where space and time are treated on equal footing. The direct spacetime formulations make manifest that the Maxwell equations are relativistically invariant. Because of this symmetry electric and magnetic field are treated on equal footing and are recognised as components of the Faraday tensor. This reduces the four Maxwell equations to two, which simplifies the equations, although we can no longer use the familiar vector formulation. In fact the Maxwell equations in the space + time formulation are not Galileo invariant and have Lorentz invariance as a hidden symmetry. This was a major source of inspiration for the development of relativity theory. To repeat: the space + time formulation is not a non-relativistic approximation and it describes the same physics by simply renaming variables. For this reason the relativistic invariant equations are usually called the Maxwell equations as well.

Each table describes one formalism.

Tensor Calculus

Formulation	Homogeneous equations	Inhomogeneous equations
Fields Minkowski space	$\partial_{[\alpha} F_{\beta\gamma]} = 0$	$\partial_\alpha F^{\alpha\beta} = \mu_0 J^\beta$
Potentials (any gauge) Minkowski space	$F_{\alpha\beta} = 2\partial_{[\alpha} A_{\beta]}$	$2\partial_\alpha \partial^{[\alpha} A^{\beta]} = \mu_0 J^\beta$
Potentials (Lorenz gauge) Minkowski space	$F_{\alpha\beta} = 2\partial_{[\alpha} A_{\beta]}\, \partial_\alpha A^\alpha = 0$	$\partial_\alpha \partial^\alpha A^\beta = \mu_0 J^\beta$
Fields Any spacetime	$\partial_{[\alpha} F_{\beta\gamma]} = \nabla_{[\alpha} F_{\beta\gamma]} = 0$	$\dfrac{1}{\sqrt{-g}}\partial_\alpha(\sqrt{-g}F^{\alpha\beta}) =$ $\nabla_\alpha F^{\alpha\beta} = \mu_0 J^\beta$
Potentials (any gauge) Any spacetime (with topological restrictions)	$F_{\alpha\beta} = 2\partial_{[\alpha} A_{\beta]}$ $= 2\nabla_{[\alpha} A_{\beta]}$	$\dfrac{2}{\sqrt{-g}}\partial_\alpha(\sqrt{-g}g^{\alpha\mu}g^{\beta\nu}\partial_{[\mu}A_{\nu]}) =$ $2\nabla_\alpha(\nabla^{[\alpha} A^{\beta]}) = \mu_0 J^\beta$
Potentials (Lorenz gauge) Any spacetime (with topological restrictions)	$F_{\alpha\beta} = 2\partial_{[\alpha} A_{\beta]}$ $= 2\nabla_{[\alpha} A_{\beta]}$ $\nabla_\alpha A^\alpha = 0$	$\nabla_\alpha \nabla^\alpha A^\beta - R^\beta{}_\alpha A^\alpha = \mu_0 J^\beta$

Differential Forms

Formulation	Homogeneous equations	Inhomogeneous equations
Fields Any spacetime	$dF = 0$	$d \star F = \mu_0 J$
Potentials (any gauge) Any spacetime (with topological restrictions)	$F = dA$	$d \star dA = \mu_0 J$
Potentials (Lorenz gauge) Any spacetime (with topological restrictions)	$F = dA$ $d \star A = 0$	$\star \Box A = \mu_0 J$

- In the tensor calculus formulation, the electromagnetic tensor $F_{\alpha\beta}$ is an antisymmetric covariant order 2 tensor; the four-potential, A_α, is a covariant vector; the current, J^α, is a vector; the square brackets, [ ], denote antisymmetrization of indices; ∂_α is the derivative with respect to the coordinate, x^α. In Minkowski space coordinates are chosen with respect to an inertial frame; $(x^\alpha) = (ct,x,y,z)$, so that the metric tensor used to raise and lower indices is $\eta_{\alpha\beta} = \text{diag}(1,-1,-1,-1)$. The d'Alembert operator on Minkowski space is $\Box = \partial_\alpha \partial^\alpha$ as in the vector formulation. In general spacetimes, the coordinate system x^α is arbitrary, the covariant derivative ∇_α, the Ricci tensor, $R_{\alpha\beta}$ and raising and lowering of indices are defined by the Lorentzian metric, $g_{\alpha\beta}$ and the d'Alembert operator is defined as $\Box = \nabla_\alpha \nabla^\alpha$. The topological restriction is that the second real cohomology group of the space vanishes. This is violated for Minkowski space with a line removed, which can model a (flat) spacetime with a point-like monopole on the complement of the line.

- In the differential form formulation on arbitrary space times, $F = F_{\alpha\beta} dx^\alpha \wedge dx^\beta$ is the electromagnetic tensor considered as a 2-form, $A = A_\alpha dx^\alpha$ is the potential 1-form, J is the current 3-form, d is the exterior derivative, and $\star$ is the Hodge star on forms defined (up to its an orientation, i.e. its sign) by the Lorentzian metric of spacetime. In the special case of 2-forms such as F, the Hodge star $\star$ depends on the metric tensor only for its local scale. This means that, as formulated, the differential form field equations are conformally invariant, but the Lorenz gauge condition breaks conformal invariance. The operator $\Box = (-\star d \star d - d \star d\star)$ is the d'Alembert–Laplace–Beltrami operator on 1-forms on an arbitrary Lorentzian spacetime. The topological condition is again that the second real cohomology group is trivial. By the isomorphism with the second de Rham cohomology this condition means that every closed 2-form is exact.

Maxwell's equations are partial differential equations that relate the electric and magnetic fields to each other and to the electric charges and currents. Often, the charges and currents are themselves dependent on the electric and magnetic fields via the Lorentz force equation and the constitutive relations. These all form a set of coupled partial differential equations which are often very difficult to solve: the solutions encompass all the diverse phenomena of classical electromagnetism. Some general remarks follow.

As for any differential equation, boundary conditions and initial conditions are necessary for a unique solution. For example, even with no charges and no currents anywhere in spacetime, there are the obvious solutions for which E and B are zero or constant, but there are also non-trivial solutions corresponding to electromagnetic waves. In some cases, Maxwell's equations are solved over the whole of space, and boundary conditions are given as asymptotic limits at infinity. In other cases, Maxwell's equations are solved in a finite region of space, with appropriate conditions on the boundary of that region, for example an artificial absorbing boundary representing the rest of the universe, or periodic boundary conditions, or walls that isolate a small region from the outside world (as with a waveguide or cavity resonator).

Jefimenko's equations (or the closely related Liénard–Wiechert potentials) are the explicit solution to Maxwell's equations for the electric and magnetic fields created by any given distribution of charges and currents. It assumes specific initial conditions to obtain the so-called "retarded solution", where the only fields present are the ones created by the charges. However, Jefimenko's equations are unhelpful in situations when the charges and currents are themselves affected by the fields they create.

Numerical methods for differential equations can be used to compute approximate solutions of Maxwell's equations when exact solutions are impossible. These include the finite element method and finite-difference time-domain method.

Overdetermination of Maxwell's Equations

Maxwell's equations seem overdetermined, in that they involve six unknowns (the three components of E and B) but eight equations (one for each of the two Gauss's laws, three vector components each for Faraday's and Ampere's laws). (The currents and charges are not unknowns, being freely specifiable subject to charge conservation.) This is related to a certain limited kind of redundancy in Maxwell's equations: It can be proven that any system satisfying Faraday's law and Ampere's law automatically also satisfies the two Gauss's laws, as long as the system's initial condition does. This explanation was first introduced by Julius Adams Stratton in 1941. Although it is possible to simply ignore the two Gauss's laws in a numerical algorithm (apart from the initial conditions), the imperfect precision of the calculations can lead to ever-increasing violations of those laws. By introducing dummy variables characterizing these violations, the four equations become not overdetermined after all. The resulting formulation can lead to more accurate algorithms that take all four laws into account.

Both identities $\nabla \cdot \nabla \times \mathbf{B} \equiv 0, \nabla \cdot \nabla \times \mathbf{E} \equiv 0$ which reduce eight equations to six independent ones, are the true reason of overdetermination.

Maxwell's Equations as the Classical Limit of QED

Maxwell's equations and the Lorentz force law (along with the rest of classical electromagnetism) are extraordinarily successful at explaining and predicting a variety of phenomena; however they are not exact, but a classical limit of quantum electrodynamics (QED).

Some observed electromagnetic phenomena are incompatible with Maxwell's equations. These include photon–photon scattering and many other phenomena related to photons or virtual photons, "nonclassical light" and quantum entanglement of electromagnetic fields. E.g. quantum cryptography cannot be described by Maxwell theory, not even approximately. The approximate nature of Maxwell's equations becomes more and more apparent when going into the extremely strong field regime or to extremely small distances.

Finally, Maxwell's equations cannot explain any phenomenon involving individual photons interacting with quantum matter, such as the photoelectric effect, Planck's law, the Duane–Hunt law, and single-photon light detectors. However, many such phenomena may be approximated using a halfway theory of quantum matter coupled to a classical electromagnetic field, either as external field or with the expected value of the charge current and density on the right hand side of Maxwell's equations.

Variations

Popular variations on the Maxwell equations as a classical theory of electromagnetic fields are relatively scarce because the standard equations have stood the test of time remarkably well.

Magnetic Monopoles

Maxwell's equations posit that there is electric charge, but no magnetic charge (*also called* magnetic monopoles), in the universe. Indeed, magnetic charge has never been observed, despite extensive searches, and may not exist. If they did exist, both Gauss's law for magnetism and Faraday's law would need to be modified, and the resulting four equations would be fully symmetric under the interchange of electric and magnetic fields.

Proper Length

Proper length or rest length refers to the length of an object in the object's rest frame.

The measurement of lengths is more complicated in the theory of relativity than in classical mechanics. In classical mechanics, lengths are measured based on the assumption that the locations of all points involved are measured simultaneously. But in the theory of relativity, the notion of simultaneity is dependent on the observer.

A different term, proper distance, provides an invariant measure whose value is the same for all observers.

Proper distance is analogous to proper time. The difference is that the proper distance is defined between two spacelike-separated events (or along a spacelike path), while the proper time is defined between two timelike-separated events (or along a timelike path).

Proper Length or Rest Length

The proper length or rest length of an object is the length of the object measured by an observer which is at rest relative to it, by applying standard measuring rods on the object. The measurement of the object's endpoints doesn't have to be simultaneous, since the endpoints are constantly at rest at the same positions in the object's rest frame, so it is independent of Δt. This length is thus given by:

$$L_0 = \Delta x.$$

However, in relatively moving frames the object's endpoints have to be measured simultaneous-ly, since they are constantly changing their position. The resulting length is shorter than the rest length, and is given by the formula for length contraction (with γ being the Lorentz factor):

$$L = \frac{L_0}{\gamma}.$$

In comparison, the invariant proper distance between two arbitrary events happening at the end-points of the same object is given by:

$$\Delta\sigma = \sqrt{\Delta x^2 - c^2 \Delta t^2}.$$

So $\Delta\sigma$ depends on Δt, whereas the object's rest length L_0 can be measured independently of Δt. It follows that $\Delta\sigma$ and L_0, measured at the endpoints of the same object, only agree with each other when the measurement events were simultaneous in the object's rest frame so that Δt is zero. As explained by Fayngold:

> "Note that the proper distance between two events is generally *not* the same as the *proper length* of an object whose end points happen to be respectively coincident with these events. Consider a solid rod of constant proper length l_0. If you are in the rest frame K_0 of the rod, and you want to measure its length, you can do it by first marking its endpoints. And it is not necessary that you mark them simultaneously in K_0. You can mark one end now (at a moment t_1) and the other end later (at a moment t_2) in K_0, and then quietly measure the distance between the marks. We can even consider such measurement as a possible opera-tional definition of proper length. From the viewpoint of the experimental physics, the re-quirement that the marks be made simultaneously is redundant for a stationary object with constant shape and size, and can in this case be dropped from such definition. Since the rod is stationary in K_0, the distance between the marks is the *proper length* of the rod regardless of the time lapse between the two markings. On the other hand, it is not the *proper distance* between the marking events if the marks are not made simultaneously in K_0."

Proper Distance between two Events in Flat Space

In special relativity, the proper distance between two spacelike-separated events is the distance between the two events, as measured in an inertial frame of reference in which the events are si-multaneous. In such a specific frame, the distance is given by:

$$\Delta\sigma = \sqrt{\Delta x^2 + \Delta y^2 + \Delta z^2},$$

where:

- Δx, Δy, and Δz are differences in the linear, orthogonal, spatial coordinates of the two events.

The definition can be given equivalently with respect to any inertial frame of reference (without requiring the events to be simultaneous in that frame) by:

$$\Delta = \sqrt{\Delta x^2 + \Delta y^2 + \Delta z^2 - c^2 \Delta t^2}$$

where:

- Δt is the difference in the temporal coordinates of the two events,

- c is the speed of light.

The two formulae are equivalent because of the invariance of spacetime intervals, and since $\Delta t = o$ exactly when the events are simultaneous in the given frame.

Two events are spacelike-separated if and only if the above formula gives a real, non-zero value for $\Delta\sigma$.

Proper Distance along a Path

The above formula for the proper distance between two events assumes that the spacetime in which the two events occur is flat. Hence, the above formula cannot in general be used in general relativity, in which curved spacetimes are considered. It is, however, possible to define the proper distance along a path in any spacetime, curved or flat. In a flat spacetime, the proper distance between two events is the proper distance along a straight path between the two events. In a curved spacetime, there may be more than one straight path (geodesic) between two events, so the proper distance along a straight path between two events would not uniquely define the proper distance between the two events.

Along an arbitrary spacelike path P, the proper distance is given in tensor syntax by the line integral:

$$L = c\int_P \sqrt{-g_{\mu\nu}dx^\mu dx^\nu},$$

where:

- $g_{\mu\nu}$ is the metric tensor for the current spacetime and coordinate mapping,

- dx^μ is the coordinate separation between neighboring events along the path P.

In the equation above, the metric tensor is assumed to use the +−−− metric signature, and is assumed to be normalized to return a time instead of a distance. The − sign in the equation should be dropped with a metric tensor that instead uses the −+++ metric signature. Also, the c should be dropped with a metric tensor that is normalized to use a distance, or that uses geometrized units.

Proper Time

In relativity, proper time along a timelike world line is defined as the time as measured by a clock following that line. It is thus independent of coordinates, and is a Lorentz scalar. The proper time interval between two events on a world line is the change in proper time. This interval is the quantity of interest, since proper time itself is fixed only up to an arbitrary additive constant, namely the setting of the clock at some event along the world line. The proper time interval between two events depends not only on the events but also the world line connecting them, and hence on the motion of the clock between the events. It is expressed as an integral over the world line. An accelerated clock will measure a smaller elapsed time between two events than that measured by a non-accelerated (inertial) clock between the same two events. The twin paradox is an example of this effect.

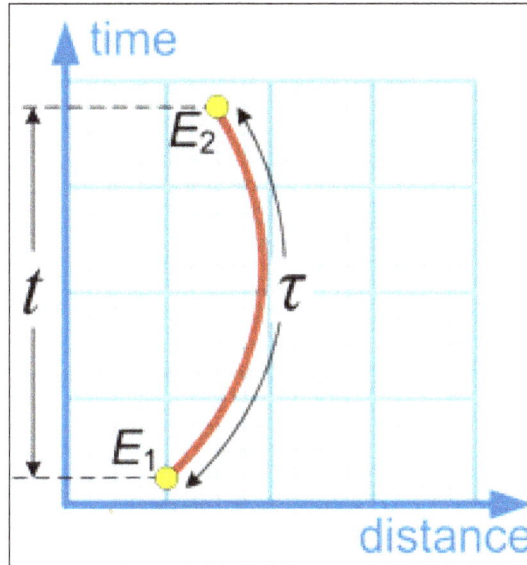

The dark blue vertical line represents an inertial observer measuring a coordinate time interval t between events E_1 and E_2. The red curve represents a clock measuring its proper time interval τ between the same two events.

In terms of four-dimensional spacetime, proper time is analogous to arc length in three-dimensional (Euclidean) space. By convention, proper time is usually represented by the Greek letter τ (tau) to distinguish it from coordinate time represented by t.

By contrast, coordinate time is the time between two events as measured by an observer using that observer's own method of assigning a time to an event. In the special case of an inertial observer in special relativity, the time is measured using the observer's clock and the observer's definition of simultaneity.

The concept of proper time was introduced by Hermann Minkowski in 1908, and is a feature of Minkowski diagrams.

Mathematical Formalism

The formal definition of proper time involves describing the path through spacetime that represents a clock, observer, or test particle, and the metric structure of that spacetime. Proper time is the pseudo-Riemannian arc length of world lines in four-dimensional spacetime. From the mathematical point of view, coordinate time is assumed to be predefined and we require an expression for proper time as a function of coordinate time. From the experimental point of view, proper time is what is measured experimentally and then coordinate time is calculated from the proper time of some inertial clocks.

Proper time can only be defined for timelike paths through spacetime which allow for the construction of an accompanying set of physical rulers and clocks. The same formalism for spacelike paths leads to a measurement of proper distance rather than proper time. For light-like paths, there exists no concept of proper time and it is undefined as the spacetime interval is identically zero. Instead an arbitrary and physically irrelevant affine parameter unrelated to time must be introduced.

In Special Relativity

Let the Minkowski metric be defined by:

$$\eta_{\mu\nu} = \begin{pmatrix} 1 & 0 & 0 & 0 \\ 0 & -1 & 0 & 0 \\ 0 & 0 & -1 & 0 \\ 0 & 0 & 0 & -1 \end{pmatrix},$$

and define:

$$(x^0, x^1, x^2, x^3) = (ct, x, y, z)$$

for arbitrary Lorentz frames.

Consider an infinitesimal interval:

$$ds^2 = c^2 dt^2 - dx^2 - dy^2 - dz^2 = \eta_{\mu\nu} dx^\mu dx^\nu,$$

expressed in any Lorentz frame and here assumed timelike, separating points on a trajectory of a particle (think clock). The same interval can be expressed in coordinates such that at each moment, the particle is *at rest*. Such a frame is called an instantaneous rest frame, denoted here by the coordinates $(c\tau, x_\tau, y_\tau, z_\tau)$ for each instants. Due to the invariance of the interval (instantaneous rest frames taken at different times are related by Lorentz transformations) one may write:

$$ds^2 = c^2 d\tau^2 - dx_\tau^2 - dy_\tau^2 - dz_\tau^2 = c^2 d\tau^2,$$

since in the instantaneous rest frame, the particle or the frame itself is at rest, i.e., $dx_\tau = dy_\tau = dz_\tau = 0$ Since the interval is assumed timelike, one may take the square root of the above expression:

$$ds = c d\tau,$$

Or

$$d\tau = \frac{ds}{c}.$$

Given this differential expression for τ, the proper time interval is defined as:

$$\Delta\tau = \int_P d\tau = \int \frac{ds}{c}.$$

Here P is the worldline from some initial event to some final event with the ordering of the events fixed by the requirement that the final event occurs later according to the clock than the initial event.

Using $ds^2 = c^2 dt^2 - dx^2 - dy^2 - dz^2 = \eta_{\mu\nu} dx^\mu dx^\nu$, and again the invariance of the interval, one may write:

$$
\begin{aligned}
\Delta\tau &= \int_P \frac{1}{c} \sqrt{\eta_{\mu\nu} dx^\mu dx^\nu} \\
&= \int_P \sqrt{dt^2 - \frac{dx^2}{c^2} - \frac{dy^2}{c^2} - \frac{dz^2}{c^2}} \\
&= \int \sqrt{1 - \frac{1}{c^2}\left[\left(\frac{dx}{dt}\right)^2 + \left(\frac{dy}{dt}\right)^2 + \left(\frac{dz}{dt}\right)^2\right]}\, dt \\
&= \int \sqrt{1 - \frac{v(t)^2}{c^2}}\, dt = \int \frac{dt}{\gamma(t)},
\end{aligned}
$$

where $v(t)$ is the coordinate speed at coordinate time t, and $x(t)$, $y(t)$, and $z(t)$ are space coordinates. The first expression is *manifestly* Lorentz invariant. They are all Lorentz invariant, since proper time and proper time intervals are coordinate-independent by definition.

If t, x, y, z, are parameterised by a parameter λ, this can be written as:

$$
\Delta\tau = \int \sqrt{\left(\frac{dt}{d\lambda}\right)^2 - \frac{1}{c^2}\left[\left(\frac{dx}{d\lambda}\right)^2 + \left(\frac{dy}{d\lambda}\right)^2 + \left(\frac{dz}{d\lambda}\right)^2\right]}\, d\lambda.
$$

If the motion of the particle is constant, the expression simplifies to:

$$
\Delta\tau = \sqrt{(\Delta t)^2 - \frac{(\Delta x)^2}{c^2} - \frac{(\Delta y)^2}{c^2} - \frac{(\Delta z)^2}{c^2}},
$$

where Δ means the change in coordinates between the initial and final events. The definition in special relativity generalizes straightforwardly to general relativity as follows below.

In General Relativity

Proper time is defined in general relativity as follows: Given a pseudo-Riemannian manifold with a local coordinates xμ and equipped with a metric tensor gμν, the proper time interval $\Delta\tau$ between two events along a timelike path P is given by the line integral:

$$
\Delta\tau = \int_P d\tau = \int_P \frac{1}{c} \sqrt{g_{\mu\nu}\, dx^\mu\, dx^\nu}.
$$

This expression is, as it should be, invariant under coordinate changes. It reduces (in appropriate coordinates) to the expression of special relativity in flat spacetime.

In the same way that coordinates can be chosen such that x^1, x^2, x^3 = const in special relativity, this can be done in general relativity too. Then, in these coordinates,

$$\Delta\tau = \int_P d\tau = \int_P \frac{1}{c}\sqrt{g_{00}}\,dx^0.$$

Examples of Special Relativity

The Twin "Paradox"

For a twin paradox scenario, let there be an observer A who moves between the A-coordinates $(0,0,0,0)$ and $(10$ years, 0, 0, $0)$ inertially. This means that A stays at $x = y = z = 0$ for 10 years of A-coordinate time. The proper time interval for A between the two events is then:

$$\Delta\tau = \sqrt{(10 \text{ years})^2} = 10 \text{ years.}$$

So being "at rest" in a special relativity coordinate system means that proper time and coordinate time are the same.

Let there now be another observer B who travels in the x direction from $(0,0,0,0)$ for 5 years of A-coordinate time at $0.866c$ to $(5$ years, 4.33 light-years, 0, $0)$. Once there, B accelerates, and travels in the other spatial direction for another 5 years of A-coordinate time to $(10$ years, 0, 0, $0)$. For each leg of the trip, the proper time interval can be calculated using A-coordinates, and is given by:

$$\Delta\tau = \sqrt{(5 \text{ years})^2 - (4.33 \text{ years})^2} = \sqrt{6.25 \text{ years}^2} = \sqrt{6.25} \text{ years} = 2.5 \text{ years.}$$

So the total proper time for observer B to go from $(0,0,0,0)$ to $(5$ years, 4.33 light-years, 0, $0)$ and then to $(10$ years, 0, 0, $0)$ is 5 years. Thus it is shown that the proper time equation incorporates the time dilation effect. In fact, for an object in a SR spacetime traveling with a velocity of v for a time ΔT, the proper time interval experienced is:

$$\Delta\tau = \sqrt{\Delta T^2 - (v_x\Delta T/c)^2 - (v_y\Delta T/c)^2 - (v_z\Delta T/c)^2} = \Delta T\sqrt{1 - v^2/c^2},$$

which is the SR time dilation formula.

The Rotating Disk

An observer rotating around another inertial observer is in an accelerated frame of reference. For such an observer, the incremental $(d\tau)$ form of the proper time equation is needed, along with a parameterized description of the path being taken, as shown below.

Let there be an observer C on a disk rotating in the xy plane at a coordinate angular rate of ω and who is at a distance of r from the center of the disk with the center of the disk at $x=y=z=0$. The path of observer C is given by $(T, r\cos(\omega T), r\sin(\omega T), 0)$ where T is the current coordinate time. When r and ω are constant, $dx = -r\omega\sin(\omega T)\,dT$ and $dy = r\omega\cos(\omega T)\,dT$. The incremental proper time formula then becomes:

$$d\tau = \sqrt{dT^2 - (r\omega/c)^2\sin^2(\omega T)\,dT^2 - (r\omega/c)^2\cos^2(\omega T)\,dT^2} = dT\sqrt{1 - \left(\frac{r\omega}{c}\right)^2}.$$

So for an observer rotating at a constant distance of r from a given point in spacetime at a constant angular rate of ω between coordinate times T_1 and T_2, the proper time experienced will be:

$$\int_{T_1}^{T_2} d\tau = (T_2 - T_1)\sqrt{1 - \left(\frac{r\omega}{c}\right)^2} = \Delta T\sqrt{1 - v^2/c^2},$$

as v = rω for a rotating observer. This result is the same as for the linear motion example, and shows the general application of the integral form of the proper time formula.

Examples of General Relativity

The difference between SR and general relativity (GR) is that in GR one can use any metric which is a solution of the Einstein field equations, not just the Minkowski metric. Because inertial motion in curved spacetimes lacks the simple expression it has in SR, the line integral form of the proper time equation must always be used.

The Rotating Disk

An appropriate coordinate conversion done against the Minkowski metric creates coordinates where an object on a rotating disk stays in the same spatial coordinate position. The new coordinates are:

$$r = \sqrt{x^2 + y^2}$$

and

$$\theta = \arctan\left(\frac{y}{x}\right) - \omega t.$$

The t and z coordinates remain unchanged. In this new coordinate system, the incremental proper time equation is:

$$d\tau = \sqrt{\left[1 - \left(\frac{r\omega}{c}\right)^2\right]dt^2 - \frac{dr^2}{c^2} - \frac{r^2\,d\theta^2}{c^2} - \frac{dz^2}{c^2} - 2\frac{r^2\omega\,dt\,d\theta}{c^2}}.$$

With r, θ, and z being constant over time, this simplifies to:

$$d\tau = dt\sqrt{1 - \left(\frac{r\omega}{c}\right)^2},$$

Now let there be an object off of the rotating disk and at inertial rest with respect to the center of the disk and at a distance of R from it. This object has a coordinate motion described by dθ = $-\omega$ dt, which describes the inertially at-rest object of counter-rotating in the view of the rotating observer. Now the proper time equation becomes:

$$d\tau = \sqrt{\left[1 - \left(\frac{R\omega}{c}\right)^2\right]dt^2 - \left(\frac{R\omega}{c}\right)^2 dt^2 + 2\left(\frac{R\omega}{c}\right)^2 dt^2} = dt.$$

So for the inertial at-rest observer, coordinate time and proper time are once again found to pass at the same rate, as expected and required for the internal self-consistency of relativity theory.

The Schwarzschild Solution – Time on the Earth

The Schwarzschild solution has an incremental proper time equation of:

$$d\tau = \sqrt{\left(1-\frac{2m}{r}\right)dt^2 - \frac{1}{c^2}\left(1-\frac{2m}{r}\right)^{-1}dr^2 - \frac{r^2}{c^2}d\phi^2 - \frac{r^2}{c^2}\sin^2(\phi)d\theta^2},$$

where,

- t is time as calibrated with a clock distant from and at inertial rest with respect to the Earth.

- r is a radial coordinate (which is effectively the distance from the Earth's center).

- ϕ is a co-latitudinal coordinate, the angular separation from the north pole in radians.

- θ is a longitudinal coordinate, analogous to the longitude on the Earth's surface but independent of the Earth's rotation. This is also given in radians.

- 1=m is the geometrized mass of the Earth, m = GM/c²:

    ○ M is the mass of the Earth.

    ○ G is the gravitational constant.

To demonstrate the use of the proper time relationship, several sub-examples involving the Earth will be used here.

For the Earth, M = 5.9742 × 10²⁴ kg, meaning that m = 4.4354 × 10⁻³ m. When standing on the north pole, we can assume $dr = d\theta = d\phi = 0$ (meaning that we are neither moving up or down or along the surface of the Earth). In this case, the Schwarzschild solution proper time equation becomes $d\tau = dt\sqrt{1-2m/r}$ Then using the polar radius of the Earth as the radial coordinate (or $r = 6,356,752$ meters), we find that:

$$d\tau = \sqrt{\left(1-1.3908\times10^{-9}\right)dt^2} = \left(1-6.9540\times10^{-10}\right)dt.$$

At the equator, the radius of the Earth is r = 6,378,137 meters. In addition, the rotation of the Earth needs to be taken into account. This imparts on an observer an angular velocity of $d\theta/dt$ of 2π divided by the sidereal period of the Earth's rotation, 86162.4 seconds. So $d\theta = 7.2923\times10^{-5}dt$ The proper time equation then produces:

$$d\tau = \sqrt{\left(1-1.3908\times10^{-9}\right)dt^2 - 2.4069\times10^{-12}dt^2} = \left(1-6.9660\times10^{-10}\right)dt.$$

From a non-relativistic point of view this should have been the same as the previous result. This example demonstrates how the proper time equation is used, even though the Earth rotates and hence is not spherically symmetric as assumed by the Schwarzschild solution. To describe the effects of rotation more accurately the Kerr metric may be used.

Relativistic Mass

The concept of mass has always been fundamental to physics. It was present in the earliest days of the subject, and its importance has only grown as physics has diversified over the centuries. Its definition goes back to Galileo and Newton, for whom mass was that property of a body that enables it to resist externally imposed changes to its motion. Newton used mass to define momentum and force vectors: he defined a body's momentum as $p = mv$ (where v is its velocity), and he defined force to be the rate of increase of the body's momentum: $F = dp/dt$. When a body's mass is constant (as it usually is, except when we are analysing the motion of e.g. a rocket), the force law becomes $F = m\, dv/dt = ma$, where a is the body's acceleration.

This definition of mass was applied in a straightforward way for almost two centuries. Then Einstein arrived on the scene and, in his theory of motion known as *special relativity*, the situation became more complicated. The above definition of mass still holds for a body at rest, and so has come to be called the body's *rest mass*, denoted m_o if we wish to stress that we're dealing with rest mass. But when the body is moving we find that its force–acceleration relationship now depends on two quantities: the body's speed, and the angle between its direction of motion and the applied force. When we relate the force to the resulting acceleration along each of three mutually perpendicular spatial axes, we find that in each of the three expressions a factor of γm_o appears, where the *gamma factor* $\gamma = (1-v^2/c^2)^{-1/2}$ occurs frequently in special relativity.

The idea of a speed-dependent mass actually dates back to Lorentz's work. His 1904 paper *Electromagnetic Phenomena in a System Moving With Any Velocity Less Than That of Light* introduced the "longitudinal" and "transverse" electromagnetic masses of the electron. With these he could write the equations of motion for an electron in an electromagnetic field in the newtonian form, provided the electron's mass was allowed to increase with its speed. Between 1905 and 1909, the relativistic theory of force, momentum, and energy was developed by Planck, Lewis, and Tolman. It turned out that a single mass dependence could be used for any acceleration, thus enabling mass to retain its independence of the body's direction of acceleration, if a speed-dependent "relativistic mass" m was understood as present in Newton's original expression $p = mv$.

So a body moving with speed v and whose momentum has magnitude p has a relativistic mass given by $m = p/v$, and (it turns out) a total energy of mc^2. A body with rest mass m_o turns out to have relativistic mass γm_o. But the definition $m = p/v$ now also neatly defines a relativistic mass for a photon: this moves with speed c and has energy E, and electromagnetic theory gives it a momentum of magnitude $p = E/c$, so it has relativistic mass $p/v = E/c^2$. The expression $m = \gamma m_o$ doesn't apply to a photon, for which γ is infinite. But on the other hand, writing $m = \gamma m_o$ won't lead to any contradictions for a photon if we define the photon's rest mass to be zero.

It seems to have been Lewis who introduced the appropriate speed dependence of mass in 1908, but the term "relativistic mass" appeared later. Relativistic mass came into common usage in the relativity texts of the early 1920s written by Pauli, Eddington, and Born. But whereas rest mass is routinely used in many areas of physics, relativistic mass is mostly restricted to the dynamics of special relativity. Because of this, a body's rest mass tends to be called simply its "mass".

The quantities that a moving observer measures as scaled by γ in special relativity are not confined to mass. Two others commonly encountered in the subject are a body's *length in the direction of motion* and its *ageing rate*, both of which get reduced by a factor of γ when measured by a passing observer. So, a ruler has a *rest length*, being the length it was given on the production line, and a *relativistic* or *contracted* length in the direction of its motion, which is the length we measure it to have as it moves past us. Likewise, a stationary clock ages normally, but when it moves it ages slowly by the gamma factor so that its "factory tick rate" is reduced by γ. Lastly, an object has a rest mass, being the mass it "came off the production line with", and a relativistic mass, being defined as above. When at rest, the object's rest mass equals its relativistic mass. When it moves, its acceleration is determined by both its relativistic mass (or its rest mass, of course) and its velocity.

The use of these γ-scaled quantities is governed only by the extent to which they are useful. While contracted length and time intervals are used—or not—insofar as they simplify special relativity analyses, relativistic mass has found itself at the centre of much debate in recent years about whether it is necessary in a physics curriculum. All physicists use rest mass, but not all physicists would have relativistic mass appear in textbooks, preferring instead always to write it in terms of rest mass when it is used (although this can't be done for photons). So, if all physicists agree that rest mass is a very fundamental concept, then why use relativistic mass at all?

When particles are moving, relativistic mass provides a very economical description that absorbs the particles' motion naturally. For example, suppose we put an object on a set of scales that are capable of measuring incredibly small increases in weight. Now heat the object. As its temperature rises causing its constituents' thermal motion to increase, the reading on the scales will increase. If we prefer to maintain the usual idea that mass is proportional to weight—assuming we don't step onto an elevator or change our home planet midway through the experiment—then it follows that the object's mass has increased. If we define mass in such a way that the object's mass does not increase as it heats up, then we will have to give up the idea that mass is proportional to weight.

Another many-particle example occurs in pre-relativistic physics, in which the centre of mass of an object is calculated by "weighting" the position vector r_i of each of its particles by their mass m_i:

$$\text{Centre of mass} = \frac{\sum_i m_i r_i}{\sum_i m_i}$$

The same expression will hold relativistically *if* each of the above masses is now a particle's *relativistic* mass. If we prefer to use only rest mass then we must replace the m_i in the above expression by $\gamma_i\, m_i$ where m_i is *rest mass*, but now the expression has lost a certain economy. Similarly, if two objects with *relativistic* masses m_1 and m_2 collide and stick together in such a way that the resulting object is at rest, then its (relativistic = rest) mass will be $m_1 + m_2$. This accords with our intuition, and intuition is mostly what good conventions are about. In contrast, a rest-mass-only analysis describes the interaction by saying that the objects have (rest) masses of M_1 and M_2, with a combined (rest) mass of $\gamma_1 M_1 + \gamma_2 M_2$. Whether our intuition has anything to gain from this new expression is not clear.

Another place where the idea of relativistic mass surfaces is when describing the *cyclotron*, a device that accelerates charged particles in circles within a constant magnetic field. The cyclotron works by applying a varying electric field to the particles, and the frequency of this variation must be tuned to the natural orbital frequency that the particles acquire as they move in the magnetic field. But in practice we find that as the particles accelerate, they begin to get out of step with the applied electric field and can no longer be accelerated further. This can be described as a consequence of their masses increasing, which changes their orbital frequency in the magnetic field.

Lastly, the energy E of an object, whether moving or at rest, is given by Einstein's famous relation $E = mc^2$, where m is its *relativistic* mass. Because, for example, the photon has no rest mass but does have relativistic mass, the use of relativistic mass makes it much easier to describe the mass changes that happen when light interacts with matter.

While relativistic mass is useful in the context of special relativity, it is rest mass that appears most often in the modern language of relativity, which centres on "invariant quantities" to build a geometrical description of relativity. Geometrical objects are useful for unifying scenarios that can be described in different coordinate systems. Because there are multiple ways of describing scenarios in relativity depending on which frame we are in, it is useful to focus on whatever invariances we can find. This is, for example, one reason why vectors (i.e. arrows) are so useful in maths and physics; everyone can use the *same* arrow to express e.g. a velocity, even though they might each quantify the arrow using different components because each observer is using different coordinates. So the reason rest mass, rest length, and proper time find their way into the tensor language of relativity is that *all* observers agree on their values. (These invariants then join with other quantities in relativity: thus, for example, the *four-force* acting on a body equals its *rest* mass times its *four-acceleration*.) Some physicists cite this view to maintain that rest mass is the only way in which mass should be understood.

As with many things, the use of relativistic mass can be a matter of taste, but it seems that at least some physicists who vehemently oppose the use of relativistic mass believe, mistakenly, that pro-relativistic mass physicists are against the idea of *rest* mass. It's not clear just why there should be this perennial confusion about preferences, and why some of those who dislike the idea of relativistic mass show such fundamentalist opposition to a choice of formalism that can never produce wrong results. The world of physics and its language is full of useful alternative notations and ways of approaching things, and different choices of notation and language can shed light on the physics involved. Selecting one of the other of relativistic versus rest mass will never lead to problems for practitioners of the subject.

A commonly heard argument against the use of relativistic mass runs as follows: "The equation $E = mc^2$ says that a body's relativistic mass is proportional to its total energy, so why should we use two terms for what is essentially the same quantity? We should just stay with energy, and use the word 'mass' to refer only to rest mass." The first difficulty with this line of reasoning is that it is quite selective; after all, it should surely rule out the use of *rest* mass as well, since within special relativity, rest mass is proportional to a body's rest energy. On that note, a second difficulty of the line of reasoning is more technical: equating energy and relativistic mass cannot be done more generally. In general relativity, it's natural to consider quantities that are conserved for a system moving on a geodesic. But $\gamma\, m$ is *not* generally conserved along geodesics. (Actually, $\gamma\, m$ is called p^t in the language of general relativity. It turns out that a closely related quantity, p_t, will be conserved along a geodesic if the metric is time independent.) Note, though, that while relativistic mass $\gamma\, m$ is not a body's total energy in general relativity, it's also not simply the source of gravity within the same theory. Finally, a third difficulty with the above commonly heard argument is that, in the interests of consistency, it should surely be applied to rule out either the "momentum density" or the "energy flux density" of light, since these also are simply related by a factor of c^2. Yet, and quite rightly, these last two terms co-exist in modern literature; no one ever suggests that either of these terms should be dropped in favour of the other, because they both have their uses and are fundamentally different quantities: a volume density and an areal density. The fact that they *can* be equated is the interesting thing physically, but one must not thereafter trivialise the physics by insisting that one of the two concepts be dropped.

So likewise can the concepts of mass and energy coexist. The above argument that $E = mc^2$ demotes mass in favour of energy—or rather, that it selectively demotes *relativistic* mass, but not *rest* mass—also neglects the very *definitions* of mass and energy. Mass is a property of a body that we have an intuitive feel for; its definition as a resistance to acceleration is very fundamental. Energy, on the other hand, is defined in physics in a technical way that involves the concept of a system's time evolution; this is not something that bears any obvious similarity to the concept of an object's resistance to being accelerated. If the concept of mass exists in some sense "prior" to that of energy, and if energy itself is defined in a different way to mass, then it does not seem reasonable to drop the idea of mass in favour of energy. Rather, $E = mc^2$ becomes an expression that tells us how much energy a given mass has; it also tells us how much a body will resist being accelerated depending on its energy content. And, perhaps best of all, it reminds us that Einstein's equation is a triumph of relating two disparate quantities—and this is one of the great aims of physics.

Another argument sometimes put forward for dropping the use of relativistic mass is that since e.g. all electrons have the same rest mass (whereas their relativistic masses depend on their speeds), then their rest mass is the only quantity able to be tabulated, and so we should discard the very idea of relativistic mass. But when we say without qualification that "the height of the Eiffel Tower is 324 metres", we clearly mean its rest length; but that doesn't mean the idea of contracted length should be discarded. Similarly, it's okay to say that the mass of an electron is about 10^{-30} kg without having to specify that we are referring to the rest mass; everyone knows we mean rest mass when we tabulate a particle's mass. That's purely a useful linguistic convention, and it does not imply that we have discarded the idea of relativistic mass, or that it should be discarded at all.

Everyone agrees that a moving train's rest mass is a fixed property inherent to it, just as its rest length is a fixed property inherent to it. And yet, strangely, many of the same physicists who insist that a moving train's mass does not scale by γ are quite happy to say that its length *does* scale by γ. There is no argument in the literature about the uses of rest length versus moving length, so why should there be any argument about the uses of rest mass versus moving mass?

Another mass concept that everyone agrees on is the idea of *reduced mass* in non-relativistic mechanics. When the mechanics of e.g. a sun–satellite system or a mass oscillating on a spring is analysed, a mass term appears that combines the two masses in a particular way. As far as the maths goes, it's *as if* we are replacing the two original bodies by two new ones: the first new body has *infinite* mass, and the second new body has a mass equal to the system's reduced mass, which has this name because it's smaller than either of the two original masses that gave rise to it. This is a fruitful way to view the original system, and it's completely standard. No one gets confused into thinking that we actually have an infinite mass and a reduced mass in our system. No one worries that the new, infinite, mass is somehow going to become a black hole, or that the reduced mass lost some of its atoms somewhere. Everyone knows the realm of applicability of the concept of reduced mass and how useful it is. Why then, do so many physicists criticise relativistic mass by squeezing it into realms where it was never intended to be used? They presumably don't do the same thing with reduced mass.

An optimistic view would hold that it's a measure of the richness of physics that focussing on different aspects of concepts like mass produces different insights: intuition in the case of relativistic mass in special relativity, and the also-intuitive notion of invariance and geometrical quantities in the case of rest mass within the tensor language of special and general relativity. The two aspects do not contradict each other, and there is room enough in the world of physics to accommodate them both.

Abandoning the use of relativistic mass is sometimes validated by quoting select physicists who are or were against the term, or by exhaustively tabulating which textbooks use the term. But real science isn't done this way. In the final analysis, the history of relativity, with its quotations from those in favour of relativistic mass and those against, has no real bearing on whether the idea itself has value. The question to ask is not whether relativistic mass is fashionable or not, or who likes the idea and who doesn't; rather, as in any area of physics notation and language, we should always ask "Is it *useful*?". And relativistic mass is certainly a useful concept. There can be little doubt that *some* of its vocal opponents even use it to gain intuition when analysing a scenario in special relativity.

Relativistic Version of F = ma

The concept of relativistic mass is neatly encapsulated in the expression $F = d(mv)/dt$, where m is relativistic mass. This says that an impulse $F\,dt$ causes an infinitesimal increase in a body's relativistic momentum mv.

Besides this definition and use of relativistic mass, we wish here to write down the relativistic version of Newton's second law, $F = ma$. In Newton's mechanics, this equation relates vectors F and a via the mass m of the object being accelerated, which is invariant in Newton's theory. Because m is just a number, in Newton's theory the force on an object is always parallel to the resulting acceleration.

The corresponding equation in special relativity is a little more complicated. It turns out that the force F is not always parallel to the acceleration a. We can express this fact using matrix notation. Let m be the *rest* mass, and v be the velocity as a column vector, whose entries are expressed as fractions of c and whose magnitude v is the speed as a fraction of c. Let v^t be the velocity as a row vector, and let 1 be the 3×3 identity matrix. As usual, set $\gamma = (1 - v^2)^{-1/2}$. The relativistic version of $F = ma$ turns out to be:

$$\mathbf{F} = \left(\mathbf{1} + \gamma^2\, \mathbf{v}\, \mathbf{v}^t\right) \gamma\, m\, \mathbf{a}$$

and

$$\mathbf{a} = \frac{\left(\mathbf{1} - \mathbf{v}\, \mathbf{v}^t\right) \mathbf{F}}{\gamma\, m}$$

So defining mass via force and acceleration isn't as simple as it was for Newton (although it *is* simple, in principle, to define the mass as relating impulse and momentum increase, as mentioned a few lines up). Nevertheless, the three components of the two expressions above share a factor of $\gamma\, m$, and the rest mass m only ever appears in both expressions accompanied by γ. The acceleration is not necessarily parallel to the force that produced it, and it's not hard to see from the above equations that it's easier to accelerate a mass sideways to its motion than it is to accelerate it in the direction of its motion. This is how relativity reproduces Lorentz's original concepts of longitudinal and transverse masses; they are actually contained in these equations. The directional dependence that the newtonian meaning of mass has now taken on is neatly contained in the matrices $1 + \gamma^2\, v\, v^t$ and $1 - v\, v^t$, and the remaining factor $\gamma\, m$ is the relativistic mass. Taking our cue from the equations like this, to isolate quantities that might prove useful, is a powerful tool in mathematical physics.

References

- What-is-special-theory-of-relativity, pure-sciences: scienceabc.com, Retrieved 12 February, 2019

- Michelson, Albert A.; Morley, Edward W. (1887). "On the Relative Motion of the Earth and the Luminiferous Ether". American Journal of Science. 34 (203): 333–345. Doi:10.2475/ajs.s3-34.203.333

- Lorentztransform, course: cv.nrao.edu, Retrieved 13 March, 2019

- Larson, Ron; Hostetler, Robert P. (2007). Elementary and Intermediate Algebra: A Combined Course, Student Support Edition (4th illudtrated ed.). Cengage Learning. P. 197. ISBN 978-0-618-75354-3

- Mass, Relativity, physics, physics: math.ucr.edu, Retrieved 16 June, 2019

- Rhodes, J. A.; Semon, M. D. (2004). "Relativistic velocity space, Wigner rotation, and Thomas precession". Am. J. Phys. 72: 93–90. Arxiv:gr-qc/0501070. Bibcode:2004amjph..72..943R. Doi:10.1119/1.1652040

Phenomena of Special Relativity 3

- **Mass–energy Equivalence**
- **Time Dilation**
- **Length Contraction**
- **Relativity of Simultaneity**
- **Relativistic Doppler Effect**
- **Thomas Precession**
- **Ladder Paradox**
- **Twin Paradox**

Some of the phenomena and concepts that are studied within special relativity are mass–energy equivalence, time dilation, length contraction, relativity of simultaneity, relativistic Doppler effect, Thomas precession, ladder paradox, twin paradox, etc. This chapter discusses in detail these phenomena and concepts related to special relativity.

Mass–energy Equivalence

Mass-energy equivalence is the famous concept in physics represented mathematically by $E=mc^2$, which states that mass and energy are one and the same. This idea was not actually put forth by Einstein, but he was the first to describe an accurate relationship for it in his theory of special relativity, where he first wrote down this famous equation. The c^2 term is a tremendously large quantity, so this means that a small amount of mass corresponds to a large amount of energy. This equation is only representative of an object at rest, so this energy is called the "rest energy" of an object. The full equation Einstein wrote down includes the energy of a moving object, but the simplified version is still profound.

The implications of such an idea are overwhelming. Mass can be created out of energy, it just takes a lot of energy to do this. In fact, the entire universe was born in the Big Bang when a whole lot of energy was turned into mass.

For example, burning a gallon of gasoline (3.78 liters) releases about 132 million joules of energy, which is enough energy to make 14 ng of mass. This is roughly the mass of a single particle of very finely ground flour. No scale in the world can detect a difference of 14 ng out of the 3 kg of mass of the gasoline.

This naturally leads to the conclusion that 'if the technology existed to just turn that gasoline into pure energy', the world's energy problems would go away. Unfortunately, this is forbidden by a deep physical law that says the total number of protons and neutrons must remain the same. Protons can become neutrons, and neutrons can become protons (and both happen with beta decay). This law is known as baryon conservation and is discussed at hyperphysics (both protons and neutrons are baryons).

In the gasoline case, a gallon of gasoline weighs about 3 kilograms (~6 pounds). The loss of a nanogram is impossible to detect with any scale, so that's all theoretical.

The strong force, weak force and electromagnetic force work together inside of a nucleus to create stable configurations of protons and neutrons. These nuclear processes make for much stronger forces than the electron recombination in the combustion of fossil fuels does. This means that the release of energy from a nuclear reaction creates enough of a mass difference to be measured. Specifically, small amounts of mass are turned into energy from the breaking up (fission) or combination (fusion) of the nuclei of atoms. Even spontaneous radioactive decay converts a bit of mass into incredible amounts of energy.

By doing so, the energy from these processes can be used to generate electricity in nuclear power plants, or as nuclear weapons, which were first deployed in World War II and have only been tested since then (nuclear weapons have since not been used as a direct attack, just very big threats). Although these reactions cannot convert the entire mass to energy, they still release tremendous amounts of energy.

The sun uses fusion of hydrogen into helium to create sunlight at an astonishing rate. The sun gives off 3.86×10^{26} W of power. That means the sun is losing 4.2 million tonnes of mass every second due to nuclear fusion.

Time Dilation

Time dilation is a difference in the elapsed time measured by two clocks, either due to them having a velocity relative to each other, or by there being a gravitational potential difference between their locations. After compensating for varying signal delays due to the changing distance between an observer and a moving clock (i.e. Doppler effect), the observer will measure the moving clock as ticking slower than a clock that is at rest in the observer's own reference frame. A clock that is close to a massive body (and which therefore is at lower gravitational potential) will record less elapsed time than a clock situated further from the said massive body (and which is at a higher gravitational potential).

These predictions of the theory of relativity have been repeatedly confirmed by experiment, and they are of practical concern, for instance in the operation of satellite navigation systems such as

GPS and Galileo. Time dilation has also been the subject of science fiction works, as it technically provides the means for forward time travel.

Time dilation explains why two working clocks will report different times after different accelerations. For example, at the ISS time goes slower, lagging 0.007 seconds behind for every six months. For GPS satellites to work, they must adjust for similar bending of spacetime to coordinate with systems on Earth.

Time dilation by the Lorentz factor was predicted by several authors at the turn of the 20th century. Joseph Larmor, at least for electrons orbiting a nucleus, wrote " individual electrons describe corresponding parts of their orbits in times shorter for the [rest] system in the ratio : $\sqrt{1 - \dfrac{v^2}{c^2}}$ Emil Cohn specifically related this formula to the rate of clocks. In the context of special relativity it was shown by Albert Einstein that this effect concerns the nature of time itself, and he was also the first to point out its reciprocity or symmetry. Subsequently, Hermann Minkowski introduced the concept of proper time which further clarified the meaning of time dilation.

Velocity Time Dilation

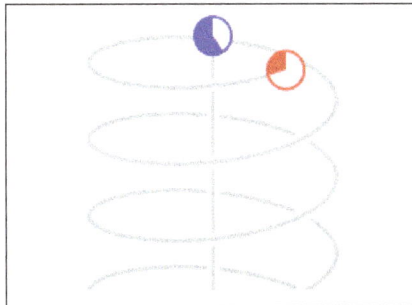

From the local frame of reference of the blue clock, the red clock, being in motion, is perceived as ticking slower.

Special relativity indicates that, for an observer in an inertial frame of reference, a clock that is moving relative to him will be measured to tick slower than a clock that is at rest in his frame of reference. This case is sometimes called special relativistic time dilation. The faster the relative velocity, the greater the time dilation between one another, with the rate of time reaching zero as one approaches the speed of light (299,792,458 m/s). This causes massless particles that travel at the speed of light to be unaffected by the passage of time.

Theoretically, time dilation would make it possible for passengers in a fast-moving vehicle to advance further into the future in a short period of their own time. For sufficiently high speeds, the effect is dramatic. For example, one year of travel might correspond to ten years on Earth. Indeed, a constant 1 g acceleration would permit humans to travel through the entire known Universe in one human lifetime.

With current technology severely limiting the velocity of space travel, however, the differences experienced in practice are minuscule: after 6 months on the International Space Station (ISS) (which orbits Earth at a speed of about 7,700 m/s) an astronaut would have aged about 0.005 seconds less than those on Earth. The cosmonauts Sergei Krikalev and Sergei Avdeyev both experienced time dilation of about 20 milliseconds compared to time that passed on Earth.

Simple Inference of Velocity Time Dilation

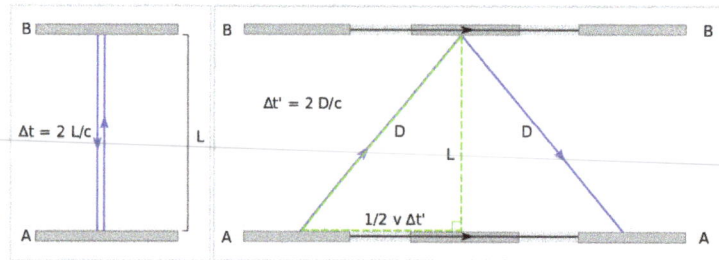

Left: Observer at rest measures time 2L/c between co-local events of light signal generation at A and arrival at A. Right: Events according to an observer moving to the left of the setup: bottom mirror A when signal is generated at time t'=0, top mirror B when signal gets reflected at time t'=D/c, bottom mirror A when signal returns at time t'=2D/c.

Time dilation can be inferred from the observed constancy of the speed of light in all reference frames dictated by the second postulate of special relativity.

This constancy of the speed of light means that, counter to intuition, speeds of material objects and light are not additive. It is not possible to make the speed of light appear greater by moving towards or away from the light source.

Consider then, a simple clock consisting of two mirrors A and B, between which a light pulse is bouncing. The separation of the mirrors is L and the clock ticks once each time the light pulse hits either of the mirrors.

In the frame in which the clock is at rest, the light pulse traces out a path of length $2L$ and the period of the clock is $2L$ divided by the speed of light:

$$\Delta t = \frac{2L}{c}.$$

From the frame of reference of a moving observer traveling at the speed v relative to the resting frame of the clock, the light pulse is seen as tracing out a longer, angled path. Keeping the speed of light constant for all inertial observers, requires a lengthening of the period of this clock from the moving observer's perspective. That is to say, in a frame moving relative to the local clock, this clock will appear to be running more slowly. Straightforward application of the Pythagorean theorem leads to the well-known prediction of special relativity.

The total time for the light pulse to trace its path is given by:

$$\Delta t' = \frac{2D}{c}.$$

The length of the half path can be calculated as a function of known quantities as:

$$D = \sqrt{\left(\frac{1}{2}v\Delta t'\right)^2 + L^2}.$$

Elimination of the variables D and L from these three equations results in:

$$\Delta t' = \frac{\Delta t}{\sqrt{1 - \dfrac{v^2}{c^2}}},$$

which expresses the fact that the moving observer's period of the clock $\Delta t'$ is longer than the period Δt in the frame of the clock itself.

Reciprocity

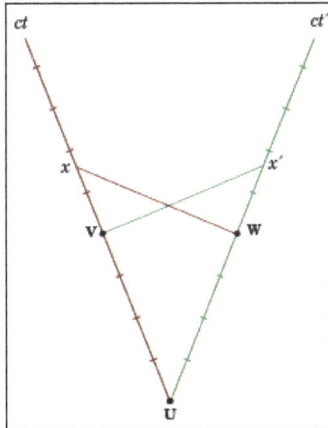

Time UV of a clock in S is shorter compared to Ux′ in S′, and time UW of a clock in S′ is shorter compared to Ux in S.

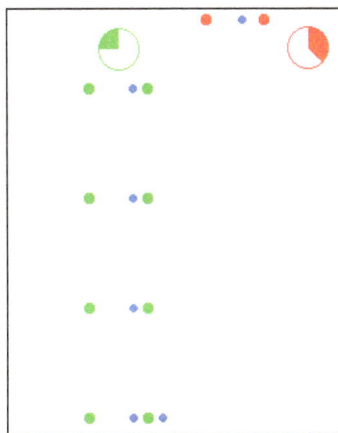

Transversal time dilation.

The blue dots represent a pulse of light. Each pair of dots with light "bouncing" between them is a clock. For each group of clocks, the other group appears to be ticking more slowly, because the moving clock's light pulse has to travel a larger distance than the stationary clock's light pulse. That is so, even though the clocks are identical and their relative motion is perfectly reciprocal.

Given a certain frame of reference, and the "stationary" observer described earlier, if a second observer accompanied the "moving" clock, each of the observers would perceive the other's clock as ticking at a *slower* rate than their own local clock, due to them both perceiving the other to be the one that's in motion relative to their own stationary frame of reference.

Common sense would dictate that, if the passage of time has slowed for a moving object, said object would observe the external world's time to be correspondingly sped up. Counterintuitively, special relativity predicts the opposite. When two observers are in motion relative to each other, each will measure the other's clock slowing down, in concordance with them being moving relative to the observer's frame of reference.

While this seems self-contradictory, a similar oddity occurs in everyday life. If two persons A and B observe each other from a distance, B will appear small to A, but at the same time A will appear small to B. Being familiar with the effects of perspective, there is no contradiction or paradox in this situation.

The reciprocity of the phenomenon also leads to the so-called twin paradox where the aging of twins, one staying on Earth and the other embarking on a space travel, is compared, and where the reciprocity suggests that both persons should have the same age when they reunite. On the contrary, at the end of the round-trip, the traveling twin will be younger than his brother on Earth. The dilemma posed by the paradox, however, can be explained by the fact that the traveling twin must markedly accelerate in at least three phases of the trip (beginning, direction change, and end), while the other will only experience negligible acceleration, due to rotation and revolution of Earth. During the acceleration phases of the space travel, time dilation is not symmetric.

Experimental Testing

Doppler Effect

- The stated purpose by Ives and Stilwell (1938, 1941) of these experiments was to verify the time dilation effect, predicted by Larmor–Lorentz ether theory, due to motion through the ether using Einstein's suggestion that Doppler effect in canal rays would provide a suitable experiment. These experiments measured the Doppler shift of the radiation emitted from cathode rays, when viewed from directly in front and from directly behind. The high and low frequencies detected were not the classically predicted values:

$$\frac{f_0}{1-v/c} \quad \text{and} \quad \frac{f_0}{1+v/c}.$$

The high and low frequencies of the radiation from the moving sources were measured as:

$$\sqrt{\frac{1+v/c}{1-v/c}}\,f_0 = \gamma\left(1+v/c\right)f_0 \quad \text{and} \quad \sqrt{\frac{1-v/c}{1+v/c}}\,f_0 = \gamma\left(1-v/c\right)f_0,$$

as deduced by Einstein from the Lorentz transformation, when the source is running slow by the Lorentz factor.

- Hasselkamp, Mondry, and Scharmann measured the Doppler shift from a source moving at right angles to the line of sight. The most general relationship between frequencies of the radiation from the moving sources is given by:

$$f_{detected} = f_{rest}\left(1-\frac{v}{c}\cos\phi\right)/\sqrt{1-v^2/c^2}$$

as deduced by Einstein. For $\phi = 90°$ (cos ϕ = 0) this reduces to $f_{detected} = f_{rest}\gamma$. This lower frequency from the moving source can be attributed to the time dilation effect and is often called the transverse Doppler effect and was predicted by relativity.

- In 2010 time dilation was observed at speeds of less than 10 meters per second using optical atomic clocks connected by 75 meters of optical fiber.

Moving Particles

- A comparison of muon lifetimes at different speeds is possible. In the laboratory, slow muons are produced; and in the atmosphere, very fast moving muons are introduced by cosmic rays. Taking the muon lifetime at rest as the laboratory value of 2.197 μs, the lifetime of a cosmic ray produced muon traveling at 98% of the speed of light is about five times longer, in agreement with observations. An example is Rossi and Hall, who compared the population of cosmic-ray-produced muons at the top of a mountain to that observed at sea level.

- The lifetime of particles produced in particle accelerators appears longer due to time dilation. In such experiments the "clock" is the time taken by processes leading to muon decay, and these processes take place in the moving muon at its own "clock rate", which is much slower than the laboratory clock. This is routinely taken into account in particle physics, and many dedicated measurements have been performed. For instance, in the muon storage ring at CERN the lifetime of muons circulating with γ = 29.327 was found to be dilated to 64.378 μs, confirming time dilation to an accuracy of 0.9 ± 0.4 parts per thousand.

Proper Time and Minkowski Diagram

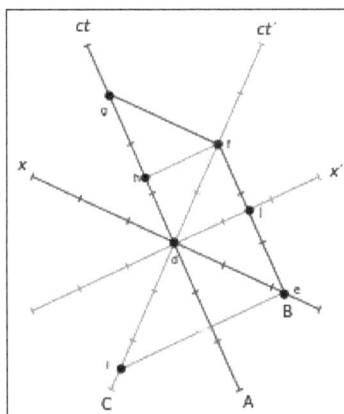

Clock C in relative motion between two synchronized clocks A and B. C meets A at *d*, and B at *f*.

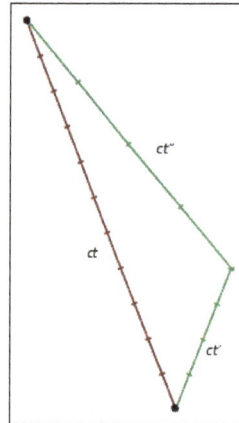

Twin paradox. One twin has to change frames, leading to different proper times in the twin's world lines.

In the Minkowski diagram from the first image on the right, clock C resting in inertial frame S′ meets clock A at d and clock B at f (both resting in S). All three clocks simultaneously start to tick in S. The worldline of A is the ct-axis, the worldline of B intersecting f is parallel to the ct-axis, and the worldline of C is the ct′-axis. All events simultaneous with d in S are on the x-axis, in S′ on the x′-axis.

The proper time between two events is indicated by a clock present at both events. It is invariant, i.e., in all inertial frames it is agreed that this time is indicated by that clock. Interval df is therefore the proper time of clock C, and is shorter with respect to the coordinate times $ef=dg$ of clocks B and A in S. Conversely, also proper time ef of B is shorter with respect to time if in S′, because event e was measured in S′ already at time i due to relativity of simultaneity, long before C started to tick.

From that it can be seen, that the proper time between two events indicated by an unaccelerated clock present at both events, compared with the synchronized coordinate time measured in all other inertial frames, is always the *minimal* time interval between those events. However, the interval between two events can also correspond to the proper time of accelerated clocks present at both events. Under all possible proper times between two events, the proper time of the unaccelerated clock is *maximal*, which is the solution to the twin paradox.

Derivation and Formulation

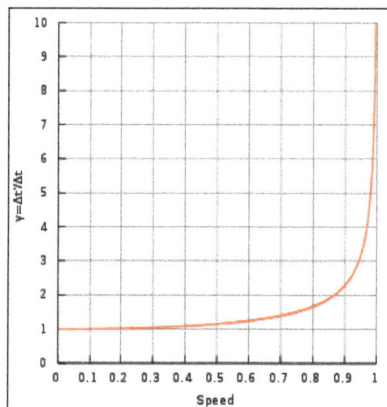

Lorentz factor as a function of speed (in natural units where $c = 1$).
Notice that for small speeds (less than 0.1), γ is approximately 1.

In addition to the light clock used above, the formula for time dilation can be more generally derived from the temporal part of the Lorentz transformation. Let there be two events at which the moving clock indicates t_a and t_b, thus:

$$t_a' = \frac{t_a - \dfrac{vx_a}{c^2}}{\sqrt{1 - \dfrac{v^2}{c^2}}}, \; t_b' = \frac{t_b - \dfrac{vx_b}{c^2}}{\sqrt{1 - \dfrac{v^2}{c^2}}}.$$

Since the clock remains at rest in its inertial frame, it follows $x_a = x_b$ thus the interval $\Delta t' = t_b' - t_a'$ is given by:

$$\Delta t' = \gamma \, \Delta t = \frac{\Delta t}{\sqrt{1 - \dfrac{v^2}{c^2}}}$$

where Δt is the time interval between *two co-local events* (i.e. happening at the same place) for an observer in some inertial frame (e.g. ticks on his clock), known as the *proper time*, $\Delta t'$ is the time interval between those same events, as measured by another observer, inertially moving with velocity v with respect to the former observer, v is the relative velocity between the observer and the moving clock, c is the speed of light, and the Lorentz factor (conventionally denoted by gamma or γ) is:

$$\gamma = \frac{1}{\sqrt{1 - \dfrac{v^2}{c^2}}}.$$

Thus the duration of the clock cycle of a moving clock is found to be increased: it is measured to be "running slow". The range of such variances in ordinary life, where v ≪ c, even considering space travel, are not great enough to produce easily detectable time dilation effects and such vanishingly small effects can be safely ignored for most purposes. It is only when an object approaches speeds on the order of 30,000 km/s (1/10 the speed of light) that time dilation becomes important.

Hyperbolic Motion

In special relativity, time dilation is most simply described in circumstances where relative velocity is unchanging. Nevertheless, the Lorentz equations allow one to calculate proper time and movement in space for the simple case of a spaceship which is applied with a force per unit mass, relative to some reference object in uniform (i.e. constant velocity) motion, equal to g throughout the period of measurement.

Let t be the time in an inertial frame subsequently called the rest frame. Let x be a spatial coordinate, and let the direction of the constant acceleration as well as the spaceship's velocity (relative to the rest frame) be parallel to the x-axis. Assuming the spaceship's position at time t = 0 being x = 0 and the velocity being v_0 and defining the following abbreviation $\gamma_0 = \dfrac{1}{\sqrt{1 - v_0^2/c^2}}$,

The following formulas hold:

Position:

$$x(t) = \frac{c^2}{g} \left(\sqrt{1 + \frac{(gt + v_0 \gamma_0)^2}{c^2}} - \gamma_0 \right).$$

Velocity:

$$v(t) = \frac{gt + v_0 \gamma_0}{\sqrt{1 + \frac{(gt + v_0 \gamma_0)^2}{c^2}}}.$$

Proper time as function of coordinate time:

$$\tau(t) = \tau_0 + \int_0^t \sqrt{1 - \left(\frac{v(t')}{c} \right)^2} \, dt'.$$

In the case where $v(0) = v_0 = 0$ and $\tau(0) = \tau 0 = 0$ the integral can be expressed as a logarithmic function or, equivalently, as an inverse hyperbolic function:

$$\tau(t) = \frac{c}{g} \ln \left(\frac{gt}{c} + \sqrt{1 + \left(\frac{gt}{c} \right)^2} \right) = \frac{c}{g} \operatorname{arsinh} \left(\frac{gt}{c} \right).$$

As functions of the proper time τ of the ship, the following formulae hold:

Position:

$$x(\tau) = \frac{c^2}{g} \left(\cosh \frac{g\tau}{c} - 1 \right).$$

Velocity:

$$v(\tau) = c \tanh \frac{g\tau}{c}.$$

Coordinate time as function of proper time:

$$t(\tau) = \frac{c}{g} \sinh \frac{g\tau}{c}.$$

Clock Hypothesis

The clock hypothesis is the assumption that the rate at which a clock is affected by time dilation does not depend on its acceleration but only on its instantaneous velocity. This is equivalent to stating that a clock moving along a path P measures the proper time, defined by:

$$d\tau = \int_P \sqrt{dt^2 - dx^2 / c^2 - dy^2 / c^2 - dz^2 / c^2}.$$

The clock hypothesis was implicitly (but not explicitly) included in Einstein's original 1905 formulation of special relativity. Since then, it has become a standard assumption and is usually included in the axioms of special relativity, especially in the light of experimental verification up to very high accelerations in particle accelerators.

Gravitational Time Dilation

Time passes more quickly further from a center of gravity, as is witnessed with massive objects (like the Earth).

Gravitational time dilation is experienced by an observer that, at a certain altitude within a gravitational potential well, finds that his local clocks measure less elapsed time than identical clocks situated at higher altitude (and which are therefore at higher gravitational potential).

Gravitational time dilation is at play e.g. for ISS astronauts. While the astronauts' relative velocity slows down their time, the reduced gravitational influence at their location speeds it up, although at a lesser degree. Also, a climber's time is theoretically passing slightly faster at the top of a mountain compared to people at sea level. It has also been calculated that due to time dilation, the core of the Earth is 2.5 years younger than the crust. "A clock used to time a full rotation of the earth will measure the day to be approximately an extra 10 ns/day longer for every km of altitude above the reference geoid." Travel to regions of space where extreme gravitational time dilation is taking place, such as near a black hole, could yield time-shifting results analogous to those of near-lightspeed space travel.

Contrarily to velocity time dilation, in which both observers measure the other as aging slower (a reciprocal effect), gravitational time dilation is not reciprocal. This means that with gravitational time dilation both observers agree that the clock nearer the center of the gravitational field is slower in rate, and they agree on the ratio of the difference.

Experimental Testing

- In 1959 Robert Pound and Glen A. Rebka measured the very slight gravitational red shift in the frequency of light emitted at a lower height, where Earth's gravitational field is relatively more intense. The results were within 10% of the predictions of general relativity. In 1964, Pound and J. L. Snider measured a result within 1% of the value predicted by gravitational time dilation.

- In 2010 gravitational time dilation was measured at the earth's surface with a height difference of only one meter, using optical atomic clocks.

Combined Effect of Velocity and Gravitational Time Dilation

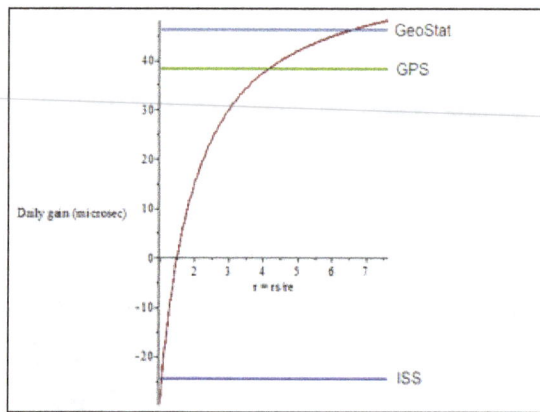

Daily time dilation (gain or loss if negative) in microseconds as a function of (circular) orbit radius $r = rs/re$, where rs is satellite orbit radius and re is the equatorial Earth radius, calculated using the Schwarzschild metric. At $r \approx 1.497$ there is no time dilation. Here the effects of motion and reduced gravity cancel.

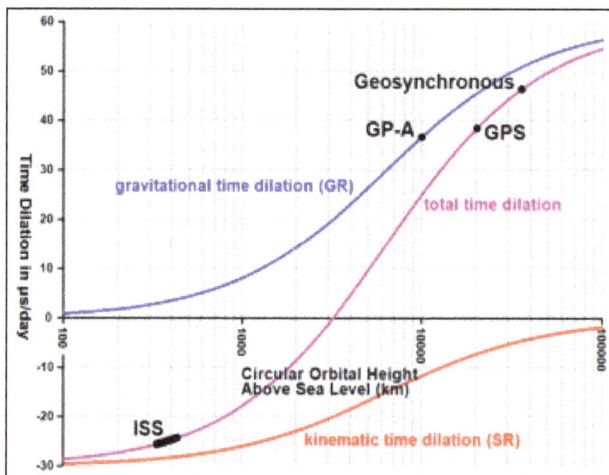

Daily time dilation over circular orbit height split into its components.

High accuracy timekeeping, low earth orbit satellite tracking, and pulsar timing are applications that require the consideration of the combined effects of mass and motion in producing time dilation. Practical examples include the International Atomic Time standard and its relationship with the Barycentric Coordinate Time standard used for interplanetary objects.

Relativistic time dilation effects for the solar system and the earth can be modeled very precisely by the Schwarzschild solution to the Einstein field equations. In the Schwarzschild metric, the interval dt_E is given by:

$$dt_E^2 = \left(1 - \frac{2GM_i}{r_i c^2}\right) dt_c^2 - \left(1 - \frac{2GM_i}{r_i c^2}\right)^{-1} \frac{dx^2 + dy^2 + dz^2}{c^2}$$

where,

- dt_E is a small increment of proper time t_E (an interval that could be recorded on an atomic clock),

- dt_c is a small increment in the coordinate t_c (coordinate time),

- dx, dy, dz are small increments in the three coordinates x, y, z of the clock's position,

- $\frac{GM_i}{r_i}$ represents the sum of the Newtonian gravitational potentials due to the masses in the neighborhood, based on their distances r_i from the clock. This sum includes any tidal potentials.

The coordinate velocity of the clock is given by:

$$v^2 = \frac{dx^2 + dy^2 + dz^2}{dt_c^2}.$$

The coordinate time t_c is the time that would be read on a hypothetical "coordinate clock" situated infinitely far from all gravitational masses $(U = 0)$, and stationary in the system of coordinates $(v = 0)$. The exact relation between the rate of proper time and the rate of coordinate time for a clock with a radial component of velocity is:

$$\frac{dt_E}{dt_c} = \sqrt{1 - \frac{2U}{c^2} - \frac{v^2}{c^2} - \left(\frac{c^2}{2U} - 1\right)^{-1} \frac{v_{||}^2}{c^2}} = \sqrt{1 - \left(\beta^2 + \beta_e^2 + \frac{\beta_{||}^2 \beta_e^2}{1 - \beta_e^2}\right)}$$

where,

- $v_{||}$ is the radial velocity,

- $v_e = \sqrt{\frac{2GM_i}{r_i}}$ is the escape velocity,

- $\beta = v/c, \beta_e = v_e/c \ \ \beta_e = v_e/c \ \ and \ \ \beta_{||} = v_{||}/c$ are velocities as a percentage of speed of light c,

- $U = \frac{GM_i}{r_i}$ is the Newtonian potential, equivalent to half of the escape velocity squared.

- The above equation is exact under the assumptions of the Schwarzschild solution. It reduces to velocity time dilation equation in the presence of motion and absence of gravity, i.e. $\beta_e = 0$ It reduces to gravitational time dilation equation in the absence of motion and presence of gravity, i.e. $\beta = 0 = \beta_{||}$.

Experimental Testing

- Hafele and Keating, in 1971, flew caesium atomic clocks east and west around the earth in commercial airliners, to compare the elapsed time against that of a clock that remained at the U.S. Naval Observatory. Two opposite effects came into play. The clocks were expected to age more quickly (show a larger elapsed time) than the reference clock, since they were in a higher (weaker) gravitational potential for most of the trip (c.f. Pound–Rebka experiment). But also, contrastingly, the moving clocks were expected to age more slowly because of the speed of their travel. From the actual flight paths of each trip, the theory predicted that the flying clocks, compared with reference clocks at the U.S. Naval Observatory, should have lost 40±23 nanoseconds during the eastward trip and should have gained 275±21 nanoseconds during the westward trip. Relative to the atomic time scale of the U.S. Naval Observatory, the flying clocks lost 59±10 nanoseconds during the eastward trip and gained 273±7 nanoseconds during the westward trip (where the error bars represent standard deviation). In 2005, the National Physical Laboratory in the United Kingdom reported their limited replication of this experiment. The NPL experiment differed from the original in that the caesium clocks were sent on a shorter trip (London–Washington, D.C. return), but the clocks were more accurate. The reported results are within 4% of the predictions of relativity, within the uncertainty of the measurements.

- The Global Positioning System can be considered a continuously operating experiment in both special and general relativity. The in-orbit clocks are corrected for both special and general relativistic time dilation effects as described above, so that (as observed from the earth's surface) they run at the same rate as clocks on the surface of the Earth.

Length Contraction

Length contraction is the phenomenon that a moving object's length is measured to be shorter than its proper length, which is the length as measured in the object's own rest frame. It is also known as Lorentz contraction or Lorentz–FitzGerald contraction (after Hendrik Lorentz and George Francis FitzGerald) and is usually only noticeable at a substantial fraction of the speed of light. Length contraction is only in the direction in which the body is travelling. For standard objects, this effect is negligible at everyday speeds, and can be ignored for all regular purposes, only becoming significant as the object approaches the speed of light relative to the observer.

Basis in Relativity

First it is necessary to carefully consider the methods for measuring the lengths of resting and moving objects. Here, "object" simply means a distance with endpoints that are always mutually at rest, *i.e.*, that are at rest in the same inertial frame of reference. If the relative velocity between an observer (or his measuring instruments) and the observed object is zero, then the proper length L_0

of the object can simply be determined by directly superposing a measuring rod. However, if the relative velocity > 0, then one can proceed as follows:

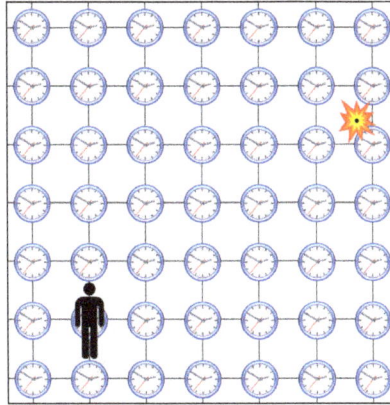

In special relativity, the observer measures events against an infinite latticework of synchronized clocks.

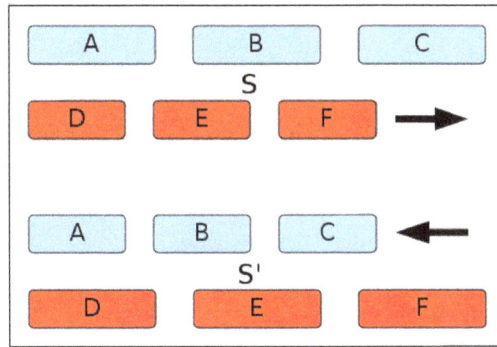

Length contraction: Three blue rods are at rest in S, and three red rods in S'. At the instant when the left ends of A and D attain the same position on the axis of x, the lengths of the rods shall be compared. In S the simultaneous positions of the left side of A and the right side of C are more distant than those of D and F. While in S' the simultaneous positions of the left side of D and the right side of F are more distant than those of A and C.

The observer installs a row of clocks that either are synchronized a) by exchanging light signals according to the Poincaré–Einstein synchronization, or b) by "slow clock transport", that is, one clock is transported along the row of clocks in the limit of vanishing transport velocity. Now, when the synchronization process is finished, the object is moved along the clock row and every clock stores the exact time when the left or the right end of the object passes by. After that, the observer only has to look at the position of a clock A that stored the time when the left end of the object was passing by, and a clock B at which the right end of the object was passing by *at the same time*. It's clear that distance AB is equal to length L of the moving object. Using this method, the definition of simultaneity is crucial for measuring the length of moving objects.

Another method is to use a clock indicating its proper time T, which is traveling from one endpoint of the rod to the other in time T as measured by clocks in the rod's rest frame. The length of the rod can be computed by multiplying its travel time by its velocity, thus $L_0 = T \cdot v$ in the rod's rest frame or $L = T_0 \cdot v$ in the clock's rest frame.

In Newtonian mechanics, simultaneity and time duration are absolute and therefore both methods lead to the equality of L and L_0. Yet in relativity theory the constancy of light velocity in all inertial frames in connection with relativity of simultaneity and time dilation destroys this equality. In the first method an observer in one frame claims to have measured the object's endpoints simultaneously, but the observers in all other inertial frames will argue that the object's endpoints were *not* measured simultaneously. In the second method, times T and T_0 are not equal due to time dilation, resulting in different lengths too.

The deviation between the measurements in all inertial frames is given by the formulas for Lorentz transformation and time dilation. It turns out that the proper length remains unchanged and always denotes the greatest length of an object, and the length of the same object measured in another inertial reference frame is shorter than the proper length. This contraction only occurs along the line of motion, and can be represented by the relation:

$$L = L_0 / \gamma(v)$$

where,

- L is the length observed by an observer in motion relative to the object.

- L_0 is the proper length (the length of the object in its rest frame).

- $\gamma(v)$ is the *Lorentz factor*, defined as:

$$\gamma(v) \equiv \frac{1}{\sqrt{1 - v^2 / c^2}}$$

where,

- v is the relative velocity between the observer and the moving object.

- c is the speed of light.

Replacing the Lorentz factor in the original formula leads to the relation:

$$L = L_0 \sqrt{1 - v^2 / c^2}$$

In this equation both L and L_0 are measured parallel to the object's line of movement. For the observer in relative movement, the length of the object is measured by subtracting the simultaneously measured distances of both ends of the object. An observer at rest observing an object travelling very close to the speed of light would observe the length of the object in the direction of motion as very near zero.

Then, at a speed of 13,400,000 m/s (30 million mph, 0.0447c) contracted length is 99.9% of the length at rest; at a speed of 42,300,000 m/s (95 million mph, 0.141c), the length is still 99%. As the magnitude of the velocity approaches the speed of light, the effect becomes prominent.

Symmetry

The principle of relativity (according to which the laws of nature must assume the same form in all inertial reference frames) requires that length contraction is symmetrical: If a rod rests in inertial frame S, it has its proper length in S and its length is contracted in S'. However, if a rod rests in S', it has its proper length in S' and its length is contracted in S. This can be vividly illustrated using symmetric Minkowski diagrams (or Loedel diagrams), because the Lorentz transformation geometrically corresponds to a rotation in four-dimensional spacetime.

Magnetic Forces

Magnetic forces are caused by relativistic contraction when electrons are moving relative to atomic nuclei. The magnetic force on a moving charge next to a current-carrying wire is a result of relativistic motion between electrons and protons.

In 1820, André-Marie Ampère showed that parallel wires having currents in the same direction attract one another. To the electrons, the wire contracts slightly, causing the protons of the opposite wire to be locally *denser*. As the electrons in the opposite wire are moving as well, they do not contract (as much). This results in an apparent local imbalance between electrons and protons; the moving electrons in one wire are attracted to the extra protons in the other. The reverse can also be considered. To the static proton's frame of reference, the electrons are moving and contracted, resulting in the same imbalance. The electron drift velocity is relatively very slow, on the order of a meter an hour but the force between an electron and proton is so enormous that even at this very slow speed the relativistic contraction causes significant effects.

This effect also applies to magnetic particles without current, with current being replaced with electron spin.

Experimental Verifications

Any observer co-moving with the observed object cannot measure the object's contraction, because he can judge himself and the object as at rest in the same inertial frame in accordance with the principle of relativity (as it was demonstrated by the Trouton–Rankine experiment). So length contraction cannot be measured in the object's rest frame, but only in a frame in which the observed object is in motion. In addition, even in such a non-co-moving frame, *direct* experimental confirmations of length contraction are hard to achieve, because at the current state of technology, objects of considerable extension cannot be accelerated to relativistic speeds. And the only objects traveling with the speed required are atomic particles, yet whose spatial extensions are too small to allow a direct measurement of contraction.

However, there are *indirect* confirmations of this effect in a non-co-moving frame:

- It was the negative result of a famous experiment, that required the introduction of length contraction: the Michelson–Morley experiment (and later also the Kennedy–Thorndike experiment). In special relativity its explanation is as follows: In its rest frame the interferometer can be regarded as at rest in accordance with the relativity principle, so the propagation time of light is the same in all directions. Although in a frame in which the interferometer is in motion, the transverse beam must traverse a longer, diagonal path with respect to the non-moving frame thus making its travel time longer, the factor by which the longitudinal beam would be delayed by taking times $L/(c-v)$ & $L/(c+v)$ for the forward and reverse trips respectively is even longer. Therefore, in the longitudinal direction the interferometer is supposed to be contracted, in order to restore the equality of both travel times in accordance with the negative experimental result(s). Thus the two-way speed of light remains constant and the round trip propagation time along perpendicular arms of the interferometer is independent of its motion & orientation.

- Given the thickness of the atmosphere as measured in Earth's reference frame, muons' extremely short lifespan shouldn't allow them to make the trip to the surface, even at the speed of light, but they do nonetheless. From the Earth reference frame, however, this is made possible only by the muon's time being slowed down by time dilation. However, in the muon's frame, the effect is explained by the atmosphere being contracted, shortening the trip.

- Heavy ions that are spherical when at rest should assume the form of "pancakes" or flat disks when traveling nearly at the speed of light. And in fact, the results obtained from particle collisions can only be explained when the increased nucleon density due to length contraction is considered.

- The ionization ability of electrically charged particles with large relative velocities is higher than expected. In pre-relativistic physics the ability should decrease at high velocities, because the time in which ionizing particles in motion can interact with the electrons of other atoms or molecules is diminished. Though in relativity, the higher-than-expected ionization ability can be explained by length contraction of the Coulomb field in frames in which the ionizing particles are moving, which increases their electrical field strength normal to the line of motion.

- In synchrotrons and free-electron lasers, relativistic electrons were injected into an undulator, so that synchrotron radiation is generated. In the proper frame of the electrons, the undulator is contracted which leads to an increased radiation frequency. Additionally, to find out the frequency as measured in the laboratory frame, one has to apply the relativistic Doppler effect. So, only with the aid of length contraction and the relativistic Doppler effect, the extremely small wavelength of undulator radiation can be explained.

Reality of Length Contraction

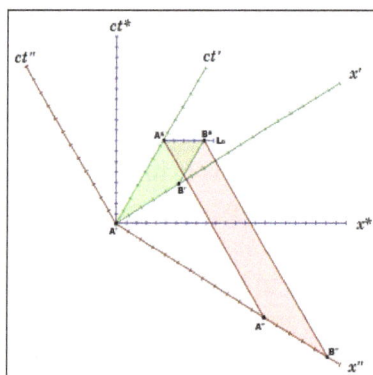

Minkowski diagram of Einstein's 1911 thought experiment on length contraction. Two rods of rest length $A'B' = A''B'' = L_0$ are moving with 0.6c in opposite directions, resulting in $A^*B^* < L_0$.

In 1911 Vladimir Varićak asserted that one sees the length contraction in an objective way, according to Lorentz, while it is "only an apparent, subjective phenomenon, caused by the manner of our clock-regulation and length-measurement", according to Einstein. Einstein published a rebuttal:

> "The author unjustifiably stated a difference of Lorentz's view and that of mine *concerning the physical facts*. The question as to whether length contraction *really* exists or not is misleading. It doesn't "really" exist, in so far as it doesn't exist for a comoving observer; though it "really" exists, *i.e.* in such a way that it could be demonstrated in principle by physical means by a non-comoving observer."

> *—Albert Einstein*

Einstein also argued in that paper, that length contraction is not simply the product of *arbitrary* definitions concerning the way clock regulations and length measurements are performed. He presented the following thought experiment: Let A'B' and A"B" be the endpoints of two rods of the same proper length L_0, as measured on x' and x" respectively. Let them move in opposite directions along the x* axis, considered at rest, at the same speed with respect to it. Endpoints A'A" then meet at point A*, and B'B" meet at point B*. Einstein pointed out that length A*B* is shorter than A'B' or A"B", which can also be demonstrated by bringing one of the rods to rest with respect to that axis.

Paradoxes

Due to superficial application of the contraction formula some paradoxes can occur. Examples are the ladder paradox and Bell's spaceship paradox. However, those paradoxes can simply be solved by a correct application of relativity of simultaneity. Another famous paradox is the Ehrenfest paradox, which proves that the concept of rigid bodies is not compatible with relativity, reducing the applicability of Born rigidity, and showing that for a co-rotating observer the geometry is in fact non-Euclidean.

Visual Effects

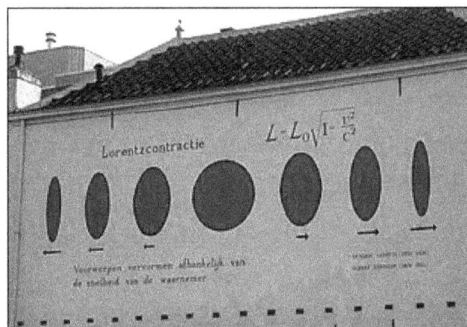

Formula on a wall in Leiden.

Length contraction refers to measurements of position made at simultaneous times according to a coordinate system. This could suggest that if one could take a picture of a fast moving object, that the image would show the object contracted in the direction of motion. However, such visual effects are completely different measurements, as such a photograph is taken from a distance, while length contraction can only directly be measured at the exact location of the object's endpoints. It was shown by several authors such as Roger Penrose and James Terrell that moving objects generally do not appear length contracted on a photograph. This result was popularized by Victor Weisskopf in a Physics Today article. For instance, for a small angular diameter, a moving sphere remains circular and is rotated. This kind of visual rotation effect is called Penrose-Terrell rotation.

Derivation

Using the Lorentz Transformation

Length contraction can be derived from the Lorentz transformation in several ways:

$$x' = \gamma\left(x - vt\right)$$
$$t' = \gamma\left(t - vx/c^2\right)$$

Known Moving Length

In an inertial reference frame S, x_1 and x_2 shall denote the endpoints of an object in motion in this frame. There, its length L was measured according to the above convention by determining the simultaneous positions of its endpoints at $t_1 = t_2$. Now, the proper length of this object in S' shall be calculated by using the Lorentz transformation. Transforming the time coordinates from S into S' results in different times, but this is not problematic, as the object is at rest in S' where it does not matter when the endpoints are measured. Therefore, the transformation of the spatial coordinates suffices, which gives:

$$x_1' = \gamma\left(x_1 - vt_1\right) \quad \text{and} \quad x_2' = \gamma\left(x_2 - vt_2\right)$$

Since $t_1 = t_2$ and by setting $L = x_2 - x_1$ and $L_0 = x_2' - x_1'$. the proper length in S' is given by:

$$L_0' = L \cdot \gamma.$$

with respect to which the measured length in S is contracted by:

$$L = L_0' / \gamma.$$

According to the relativity principle, objects that are at rest in S have to be contracted in S' as well. By exchanging the above signs and primes symmetrically, it follows:

$$L_0 = L' \cdot \gamma.$$

Thus the contracted length as measured in S' is given by:

$$L' = L_0 / \gamma.$$

Known Proper Length

Conversely, if the object rests in S and its proper length is known, the simultaneity of the measurements at the object's endpoints has to be considered in another frame S', as the object constantly changes its position there. Therefore, both spatial and temporal coordinates must be transformed:

$$x_1' = \gamma(x_1 - vt_1) \quad \text{and} \quad x_2' = \gamma(x_2 - vt_2)$$
$$t_1' = \gamma(t_1 - vx_1/c^2) \quad \text{and} \quad t_2' = \gamma(t_2 - vx_2/c^2)$$

With $t_1 = t_2$ and $L_0 = x_2 - x_1$ this results in non-simultaneous differences:

$$\Delta x' \equiv x_2' - x_1' = \gamma L_0$$
$$\Delta t' \equiv t_2' - t_1' = -\gamma v L_0 / c^2$$

In order to obtain the simultaneous positions of both endpoints, the second endpoint must be advanced by $-\Delta t$ with the speed $-v$ of S relative to S'. To obtain the length L', the quantity $(-v)\cdot(-\Delta t)$ must therefore be added to $\Delta x'$:

$$L' = \Delta x' + v\Delta t'$$
$$= \gamma L_0 - \gamma v^2 L_0 / c^2$$
$$= L_0 / \gamma$$

So the moving length in S' is contracted. Likewise, the preceding calculation gives a symmetric result for an object at rest in S':

$$L = L_0' / \gamma$$

Using Time Dilation

Length contraction can also be derived from time dilation, according to which the rate of a single "moving" clock (indicating its proper time T_0) is lower with respect to two synchronized "resting" clocks (indicating T). Time dilation was experimentally confirmed multiple times, and is represented by the relation:

$$T = T_0 \cdot \gamma$$

Suppose a rod of proper length L_0 at rest in S and a clock at rest in S' are moving along each other with speed v. Since, according to the principle of relativity, the magnitude of relative velocity is the same in either reference frame, the respective travel times of the clock between the rod's endpoints are given by $T = L_0/v$ in S and $T_0' = L'/v$ in S', thus $L_0 = Tv$ and $L' = T_0'v$. By inserting the time dilation formula, the ratio between those lengths is:

$$\frac{L'}{L_0} = \frac{T_0'v}{Tv} = 1/\gamma.$$

Therefore, the length measured in S' is given by:

$$L' = L_0 / \gamma$$

So since the clock's travel time across the rod is longer in S than in S' (time dilation in S'), the rod's length is also longer in S' than in S' (length contraction in S'). Likewise, if the clock were at rest in S' and the rod in S', the above procedure would give:

$$L = L'_0 / \gamma$$

Geometrical Considerations

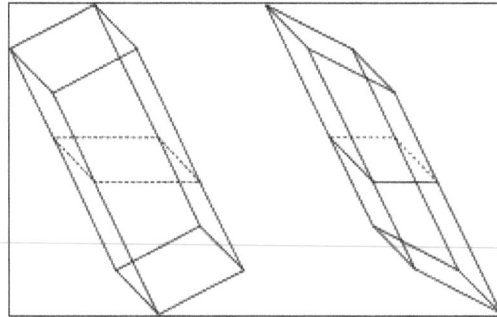

Cuboids in Euclidean and Minkowski spacetime.

A rotated cuboid in three-dimensional euclidean space E^3. The cross section is longer in the direction of the rotation than it was before the rotation. The world slab of a moving thin plate in Minkowski spacetime (with one spatial dimension suppressed) $E^{1,2}$, which is a boosted cuboid. The cross section is thinner in the direction of the boost than it was before the boost. In both cases, the transverse directions are unaffected and the three planes meeting at each corner of the cuboids are mutually orthogonal.

Additional geometrical considerations show, that length contraction can be regarded as a trigonometric phenomenon, with analogy to parallel slices through a cuboid before and after a *rotation* in E^3. This is the Euclidean analog of *boosting* a cuboid in $E^{1,2}$. In the latter case, however, we can interpret the boosted cuboid as the *world slab* of a moving plate.

In special relativity, Poincaré transformations are a class of affine transformations which can be characterized as the transformations between alternative Cartesian coordinate charts on Minkowski spacetime corresponding to alternative states of inertial motion (and different choices of an origin). Lorentz transformations are Poincaré transformations which are linear transformations (preserve the origin). Lorentz transformations play the same role in Minkowski geometry (the Lorentz group forms the *isotropy group* of the self-isometries of the spacetime) which are played by rotations in euclidean geometry. Indeed, special relativity largely comes down to studying a kind of noneuclidean trigonometry in Minkowski spacetime, as suggested by the following table:

Three Plane Trigonometries

Trigonometry	Circular	Parabolic	Hyperbolic
Kleinian Geometry	Euclidean plane	Galilean plane	Minkowski plane
Symbol	E^2	$E^{0,1}$	$E^{1,1}$
Quadratic form	positive definite	degenerate	non-degenerate but indefinite
Isometry group	E(2)	E(0,1)	E(1,1)

Isotropy group	SO(2)	SO(0,1)	SO(1,1)
type of isotropy	rotations	shears	boosts
Algebra over R	complex numbers	dual numbers	split-complex numbers
ε^2	-1	0	1
Spacetime interpretation	none	Newtonian spacetime	Minkowski spacetime
slope	$\tan \varphi = m$	$\tan p \; \varphi = u$	$\tanh \varphi = v$
"cosine"	$\cos \varphi = (1+m^2)^{-1/2}$	$\cos p \; \varphi = 1$	$\cosh \varphi = (1-v^2)^{-1/2}$
"sine"	$\sin \varphi = m \, (1+m^2)^{-1/2}$	$\sin p \; \varphi = u$	$\sinh \varphi = v \, (1-v^2)^{-1/2}$
"secant"	$\sec \varphi = (1+m^2)^{1/2}$	$\sec p \; \varphi = 1$	$\operatorname{sech} \varphi = (1-v^2)^{1/2}$
"cosecant"	$\csc \varphi = m^{-1} \, (1+m^2)^{1/2}$	$\csc p \; \varphi = u^{-1}$	$\operatorname{csch} \varphi = v^{-1} \, (1-v^2)^{1/2}$

Relativity of Simultaneity

Time intervals depend on who observes them Intuitively, it seems that the time for a process, such as the elapsed time for a foot race, should be the same for all observers. In everyday experiences, disagreements over elapsed time have to do with the accuracy of measuring time. No one would be likely to argue that the actual time interval was different for the moving runner and for the stationary clock displayed. Carefully considering just how time is measured, however, shows that elapsed time does depends on the relative motion of an observer with respect to the process being measured.

Figure : Elapsed time for a foot race is the same for all observers, but at relativistic speeds, elapsed time depends on the motion of the observer relative to the location where the process being timed occurs.

Consider how we measure elapsed time. If we use a stopwatch, for example, how do we know when to start and stop the watch? One method is to use the arrival of light from the event. For example, if you're in a moving car and observe the light arriving from a traffic signal change from green to red, you know it's time to step on the brake pedal. The timing is more accurate if some sort of electronic detection is used, avoiding human reaction times and other complications.

Now suppose two observers use this method to measure the time interval between two flashes of light from flash lamps that are a distance apart. An observer A is seated midway on a rail car with two flash lamps at opposite sides equidistant from her. A pulse of light is emitted from each flash lamp and moves toward observer A, shown in frame (a) of the figure. The rail car is moving rapidly in the direction indicated by the velocity vector in the diagram. An observer *B* standing on the platform is facing the rail car as it passes and observes both flashes of light reach him simultaneously, as shown in frame (c). He measures the distances from where he saw the pulses originate, finds them equal, and concludes that the pulses were emitted simultaneously.

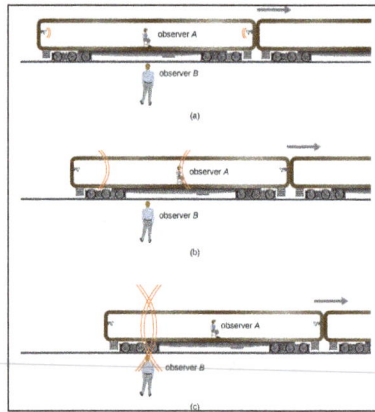

(a) Two pulses of light are emitted simultaneously relative to observer B. (c) The pulses reach observer B's position simultaneously. (b) Because of A's motion, she sees the pulse from the right first and concludes the bulbs did not flash simultaneously. Both conclusions are correct.

However, because of Observer *A*'s motion, the pulse from the right of the railcar, from the direction the car is moving, reaches her before the pulse from the left, as shown in frame (b). She also measures the distances from within her frame of reference, finds them equal, and concludes that the pulses were not emitted simultaneously.

The two observers reach conflicting conclusions about whether the two events at well-separated locations were simultaneous. Both frames of reference are valid, and both conclusions are valid. Whether two events at separate locations are simultaneous depends on the motion of the observer relative to the locations of the events.

Here, the relative velocity between observers affects whether two events a distance apart are observed to be simultaneous. Simultaneity is not absolute. We might have guessed (incorrectly) that if light is emitted simultaneously, then two observers halfway between the sources would see the flashes simultaneously. But careful analysis shows this cannot be the case if the speed of light is the same in all inertial frames.

This type of thought experiment shows that seemingly obvious conclusions must be changed to agree with the postulates of relativity. The validity of thought experiments can only be determined by actual observation, and careful experiments have repeatedly confirmed Einstein's theory of relativity.

Relativistic Doppler Effect

The relativistic Doppler effect is the change in frequency (and wavelength) of light, caused by the relative motion of the source and the observer (as in the classical Doppler effect), when taking into account effects described by the special theory of relativity.

The relativistic Doppler effect is different from the non-relativistic Doppler effect as the equations include the time dilation effect of special relativity and do not involve the medium of propagation as a reference point. They describe the total difference in observed frequencies and possess the required Lorentz symmetry.

Astronomers know of three sources of redshift/blueshift: Doppler shifts; gravitational redshifts (due to light exiting a gravitational field); and cosmological expansion (where space itself stretches).

Scenario	Formula	Notes
Relativistic longitudinal Doppler effect	$\dfrac{\lambda_r}{\lambda_s} = \dfrac{f_s}{f_r} = \sqrt{\dfrac{1+\beta}{1-\beta}}$	
Transverse Doppler effect, geometric closest approach	$f_r = \gamma f_s$	Blueshift
Transverse Doppler effect, visual closest approach	$f_r = \dfrac{f_s}{\gamma}$	Blueshift
TDE, receiver in circular motion around source	$f_r = \gamma f_s$	Blueshift
TDE, source in circular motion around receiver	$f_r = \dfrac{f_s}{\gamma}$	Blueshift
TDE, source and receiver in circular motion around common center	$\dfrac{v'}{v} = \left(\dfrac{1-R^2\omega^2}{1-R'^2\omega^2}\right)^{1/2}$	No Doppler shift when $R = R'$
Motion in arbitrary direction measured in receiver frame	$f_r = \dfrac{f_s}{\gamma\left(1+\beta\cos\theta_r\right)}$	
Motion in arbitrary direction measured in source frame	$f_r = \gamma\left(1-\beta\cos\theta_s\right)f_s$	

Derivation

Relativistic Longitudinal Doppler Effect

Relativistic Doppler shift for the longitudinal case, with source and receiver moving directly towards or away from each other, is often derived as if it were the classical phenomenon, but modified by the addition of a time dilation term. This is the approach employed in first-year physics or mechanics textbooks such as those by Feynman or Morin.

Following this approach towards deriving the relativistic longitudinal Doppler effect, assume the receiver and the source are moving *away* from each other with a relative speed v as measured by an observer on the receiver or the source (The sign convention adopted here is that v is *negative* if the receiver and the source are moving *towards* each other).

Suppose one wavefront arrives at the receiver. The next wavefront is then at a distance $\lambda_s = c / f_s$ away from the receiver (where λ_s is the wavelength, f_s is the frequency of the waves that the source emits, and c, is the speed of light).

The wavefront moves with speed c, but at the same time the receiver moves away with speed v during a time $t_s = 1/ f_s = \lambda_s / c, so,$

$$\lambda_s + vt_{r,s} = ct_{r,s} \Leftrightarrow \lambda_s = ct_{r,s}(1 - v/c) \Leftrightarrow t_{r,s} = \frac{1}{f_s(1-\beta)},$$

where $\beta = v/c$ is the speed of the receiver in terms of the speed of light, and where $t_{r,s}$ is the period of light waves impinging on the receiver, as observed in the frame of the source. The corresponding frequency $f_{r,s}$ is:

$$f_{r,s} = 1/t_{r,s} = f_s(1-\beta).$$

Thus far, the equations have been identical to those of the classical Doppler effect with a stationary source and a moving receiver.

However, due to relativistic effects, clocks on the receiver are time dilated relative to clocks at the source: $t_r = t_{r,s}/\gamma$ where $\gamma = 1/\sqrt{1-\beta^2}$ is the Lorentz factor. In order to know which time is dilated, we recall that $t_{r,s}$ is the time in the frame in which the source is at rest. The receiver will measure the received frequency to be:

$$f_r = f_{r,s}\gamma = \frac{1-\beta}{\sqrt{1-\beta^2}} f_s = \sqrt{\frac{1-\beta}{1+\beta}} f_s.$$

The ratio:

$$\frac{f_s}{f_r} = \sqrt{\frac{1+\beta}{1-\beta}}$$

is called the Doppler factor of the source relative to the receiver.

The corresponding wavelengths are related by:

$$\frac{\lambda_r}{\lambda_s} = \frac{f_s}{f_r} = \sqrt{\frac{1+\beta}{1-\beta}},$$

Identical expressions for relativistic Doppler shift are obtained when performing the analysis in the reference frame of the receiver with a moving source. This matches up with the expectations of the principle of relativity, which dictates that the result can not depend on which object is considered to be the one at rest. In contrast, the classic nonrelativistic Doppler effect is dependent on whether it is the source or the receiver that is stationary with respect to the medium.

Transverse Doppler Effect

Suppose that a source and a receiver are both approaching each other in uniform inertial motion along paths that do not collide. The *transverse Doppler effect* (TDE) may refer to (a) the nominal blueshift predicted by special relativity that occurs when the emitter and receiver are at their points of closest approach; or (b) the nominal redshift predicted by special relativity when the receiver sees the emitter as being at its closest approach. The transverse Doppler effect is one of the main novel predictions of the special theory of relativity.

Whether a scientific report describes TDE as being a redshift or blueshift depends on the particulars of the experimental arrangement being related. For example, Einstein's original description of the TDE in 1907 described an experimenter looking at the center (nearest point) of a beam of "canal rays" (a beam of positive ions that is created by certain types of gas-discharge tubes). According to special relativity, the moving ions' emitted frequency would be reduced by the Lorentz factor, so that the received frequency would be reduced (redshifted) by the same factor.

On the other hand, Kündig described an experiment where a Mössbauer absorber was spun in a rapid circular path around a central Mössbauer emitter. As explained below, this experimental arrangement resulted in Kündig's measurement of a blueshift.

Source and Receiver are at their Points of Closest Approach

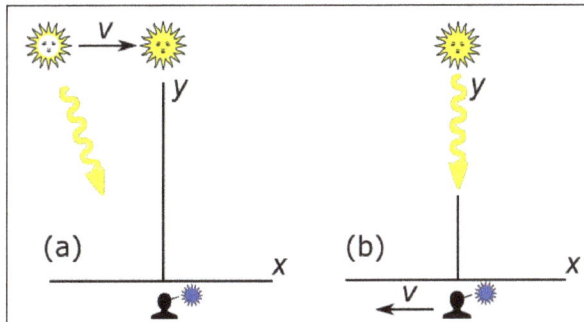

Source and receiver are at their points of closest approach. (a) Analysis in the frame of the receiver. (b) Analysis in the frame of the source.

In this scenario, the point of closest approach is frame-independent and represents the moment where there is no change in distance versus time. Figure demonstrates that the ease of analyzing this scenario depends on the frame in which it is analyzed.

- If we analyze the scenario in the frame of the receiver, we find that the analysis is more complicated than it should be. The apparent position of a celestial object is displaced from its true position (or geometric position) because of the object's motion during the time it takes its light to reach an observer. The source would be time-dilated relative to the receiver, but the redshift implied by this time dilation would be offset by a blueshift due to the longitudinal component of the relative motion between the receiver and the apparent position of the source.

- It is much easier if, instead, we analyze the scenario from the frame of the source. An observer situated at the source knows, from the problem statement, that the receiver is at its closest point to him. That means that the receiver has no longitudinal component of motion to complicate the analysis. (i.e. dr/dt = 0 where r is the distance between receiver and source) Since the receiver's clocks are time-dilated relative to the source, the light that the receiver receives is blue-shifted by a factor of gamma. In other words,

$$f_r = \gamma f_s$$

Receiver *Sees* the Source as Being at its Closest Point

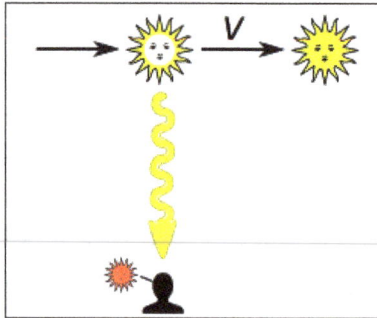

Transverse Doppler shift for the scenario where the receiver *sees* the source as being at its closest point.

This scenario is equivalent to the receiver looking at a direct right angle to the path of the source. The analysis of this scenario is best conducted from the frame of the receiver. Figure shows the receiver being illuminated by light from when the source was closest to the receiver, even though the source has moved on. Because the source's clock is time dilated as measured in the frame of the receiver, and because there is no longitudinal component of its motion, the light from the source, emitted from this closest point, is redshifted with frequency:

$$f_r = \frac{f_s}{\gamma}$$

In the literature, most reports of transverse Doppler shift analyze the effect in terms of the receiver pointed at direct right angles to the path of the source, thus seeing the source as being at its closest point and observing a redshift.

Point of Null Frequency Shift

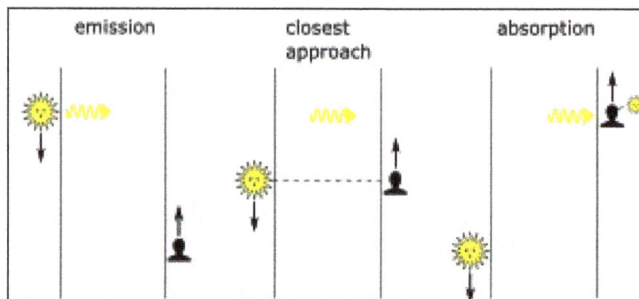

Null frequency shift occurs for a pulse that travels the shortest distance from source to receiver.

Given that, in the case where the inertially moving source and receiver are geometrically at their nearest approach to each other, the receiver observes a blueshift, whereas in the case where the receiver *sees* the source as being at its closest point, the receiver observes a redshift, there obviously must exist a point where blueshift changes to a redshift. In figure, the signal travels perpendicular-ly to the receiver path and is blueshifted. In figure, the signal travels perpendicularly to the source path and is redshifted.

As seen in figure, null frequency shift occurs for a pulse that travels the shortest distance from source to receiver. When viewed in the frame where source and receiver have the same speed, this pulse is emitted perpendicularly to the source's path and is received perpendicularly to the receiver's path. The pulse is emitted slightly before the point of closest approach, and it is received slightly after.

One Object in Circular Motion around the other

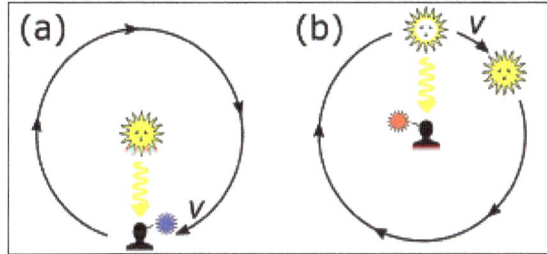

Transverse Doppler effect for two scenarios: (a) receiver moving in a circle around the source; (b) source moving in a circle around the receiver.

Figure illustrates two variants of this scenario. Both variants can be analyzed using simple time dilation arguments. Figure is essentially equivalent to the scenario described in figure, and the receiver observes light from the source as being blueshifted by a factor of γ. Figure is essentially equivalent to the scenario described in figure, and the light is redshifted.

The only seeming complication is that the orbiting objects are in accelerated motion. An accelerat-ed particle does not have an inertial frame in which it is always at rest. However, an inertial frame can always be found which is momentarily comoving with the particle. This frame, the *momentar-ily comoving reference frame* (MCRF), enables application of special relativity to the analysis of accelerated particles. If an inertial observer looks at an accelerating clock, only the clock's instan-taneous speed is important when computing time dilation.

The converse, however, is not true. The analysis of scenarios where *both* objects are in accelerated motion requires a somewhat more sophisticated analysis. Not understanding this point has led to confusion and misunderstanding.

Source and Receiver Both in Circular Motion around a Common Center

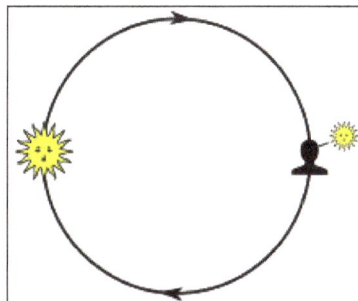

Source and receiver are placed on opposite ends of a rotor, equidistant from the center.

Suppose source and receiver are located on opposite ends of a spinning rotor, as illustrated in figure. Kinematic arguments (special relativity) and arguments based on noting that there is no difference in potential between source and receiver in the pseudogravitational field of the rotor (general relativity) both lead to the conclusion that there should be no Doppler shift between source and receiver.

In 1961, Champeney and Moon conducted a Mössbauer rotor experiment testing exactly this scenario, and found that the Mössbauer absorption process was unaffected by rotation. They concluded that their findings supported special relativity.

This conclusion generated some controversy. A certain persistent critic of relativity maintained that, although the experiment was consistent with general relativity, it refuted special relativity, his point being that since the emitter and absorber were in uniform relative motion, special relativity demanded that a Doppler shift be observed. The fallacy with this critic's argument was, as demonstrated in section Point of null frequency shift, that it is simply not true that a Doppler shift must always be observed between two frames in uniform relative motion. Furthermore, as demonstrated in section Source and receiver are at their points of closest approach, the difficulty of analyzing a relativistic scenario often depends on the choice of reference frame. Attempting to analyze the scenario in the frame of the receiver involves much tedious algebra. It is much easier, almost trivial, to establish the lack of Doppler shift between emitter and absorber in the laboratory frame.

As a matter of fact, however, Champeney and Moon's experiment said nothing either pro or con about special relativity. Because of the symmetry of the setup, it turns out that virtually *any* conceivable theory of the Doppler shift between frames in uniform inertial motion must yield a null result in this experiment.

Rather than being equidistant from the center, suppose the emitter and absorber were at differing distances from the rotor's center. For an emitter at radius R' and the absorber at radius R *anywhere* on the rotor, the ratio of the emitter frequency, v' and the absorber frequency, v is given by:

$$\frac{v'}{v} = \left(\frac{1 - R^2\omega^2}{1 - R'^2\omega^2} \right)^{1/2}$$

where ω is the angular velocity of the rotor. The source and emitter do not have to be 180° apart, but can be at any angle with respect to the center.

Motion in an Arbitrary Direction

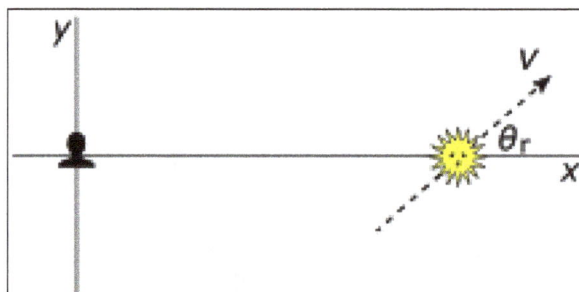

Doppler shift with source moving at an arbitrary angle with respect to the line between source and receiver.

The analysis can be extended in a straightforward fashion to calculate the Doppler shift for the case where the inertial motions of the source and receiver are at any specified angle. Figure presents the scenario from the frame of the receiver, with the source moving at speed v at an angle θ_r measured in the frame of the receiver. The radial component of the source's motion along the line of sight is equal to $v\cos\theta_r$.

The equation below can be interpreted as the classical Doppler shift for a stationary and moving source modified by the Lorentz factor γ:

$$f_r = \frac{f_s}{\gamma(1+\beta\cos\theta_r)}.$$

In the case when $\theta_r = 90°$ and $\cos\theta_r = 0$, one obtains the transverse Doppler effect:

$$f_r = \frac{f_s}{\gamma}.$$

In his 1905 paper on special relativity, Einstein obtained a somewhat different looking equation for the Doppler shift equation. After changing the variable names in Einstein's equation to be consistent with those used here, his equation reads:

$$f_r = \gamma(1-\beta\cos\theta_s)f_s.$$

The differences stem from the fact that Einstein evaluated the angle θ_s with respect to the source rest frame rather than the receiver rest frame. θ_r is not equal to θ_s because of the effect of relativistic aberration. The relativistic aberration equation is:

$$\cos\theta_r = \frac{\cos\theta_s - \beta}{1 - \beta\cos\theta_s}$$

Substituting the relativistic aberration equation $\cos\theta_r = \dfrac{\cos\theta_s - \beta}{1 - \beta\cos\theta_s}$ into $f_r = \dfrac{f_s}{\gamma(1+\beta\cos\theta_r)}$

yields $f_r = \gamma(1-\beta\cos\theta_s)f_s$, demonstrating the consistency of these alternate equations for the Doppler shift.

Setting $\theta_r = 0$ in $f_r = \dfrac{f_s}{\gamma(1+\beta\cos\theta_r)}$ or $\theta_s = 0$ in $f_r = \gamma(1-\beta\cos\theta_s)f_s$ yields:

$$f_r = f_{r,s}\gamma = \frac{1-\beta}{\sqrt{1-\beta^2}}f_s = \sqrt{\frac{1-\beta}{1+\beta}}f_s,$$ the expression for relativistic longitudinal Doppler shift.

A four-vector approach to deriving these results may be found in Landau and Lifshitz.

Visualization

Figure helps us understand, in a rough qualitative sense, how the relativistic Doppler effect and relativistic aberration differ from the non-relativistic Doppler effect and non-relativistic aberration of light. Assume that the observer is uniformly surrounded in all directions by yellow stars emitting monochromatic light of 570 nm. The arrows in each diagram represent the observer's velocity vector relative to its surroundings, with a magnitude of 0.89 c.

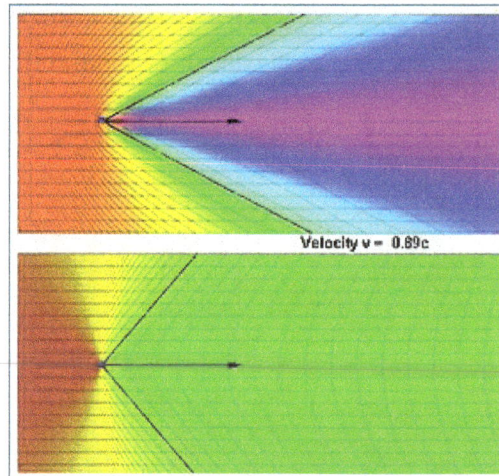

Comparison of the relativistic Doppler effect (top) with the
non-relativistic effect (bottom).

- In the non-relativistic case, the light ahead of the observer is blueshifted to a wavelength of 300 nm in the medium ultraviolet, while light behind the observer is redshifted to 5200 nm in the intermediate infrared. Because of the aberration of light, objects formerly at right angles to the observer appear shifted forwards by 42°.

- In the relativistic case, the light ahead of the observer is blueshifted to a wavelength of 137 nm in the far ultraviolet, while light behind the observer is redshifted to 2400 nm in the short wavelength infrared. Because of the relativistic aberration of light, objects formerly at right angles to the observer appear shifted forwards by 63°.

- In both cases, the monochromatic stars ahead of and behind the observer are Doppler-shifted towards invisible wavelengths. If, however, the observer had eyes that could see into the ultraviolet and infrared, he would see the stars ahead of him as brighter and more closely clustered together than the stars behind, but the stars would be far brighter and far more concentrated in the relativistic case.

Real stars are not monochromatic, but emit a range of wavelengths approximating a black body distribution. It is not necessarily true that stars ahead of the observer would show a bluer color. This is because the whole spectral energy distribution is shifted. At the same time that visible light is blueshifted into invisible ultraviolet wavelengths, infrared light is blueshifted into the visible range. Precisely what changes in the colors one sees depends on the physiology of the human eye and on the spectral characteristics of the light sources being observed.

Doppler Effect on Intensity

The Doppler effect (with arbitrary direction) also modifies the perceived source intensity: this can be expressed concisely by the fact that source strength divided by the cube of the frequency is a

Lorentz invariant. This implies that the total radiant intensity (summing over all frequencies) is multiplied by the fourth power of the Doppler factor for frequency.

As a consequence, since Planck's law describes the black-body radiation as having a spectral intensity in frequency proportional to $v^3 / \left(e^{hv/kT} - 1 \right)$ (where T is the source temperature and v the frequency), we can draw the conclusion that a black body spectrum seen through a Doppler shift (with arbitrary direction) is still a black body spectrum with a temperature multiplied by the same Doppler factor as frequency.

This result provides one of the pieces of evidence that serves to distinguish the Big Bang theory from alternative theories proposed to explain the cosmological redshift.

Experimental Verification

Since the transverse Doppler effect is one of the main novel predictions of the special theory of relativity, the detection and precise quantification of this effect has been an important goal of experiments attempting to validate special relativity.

Ives and Stilwell-type Measurements

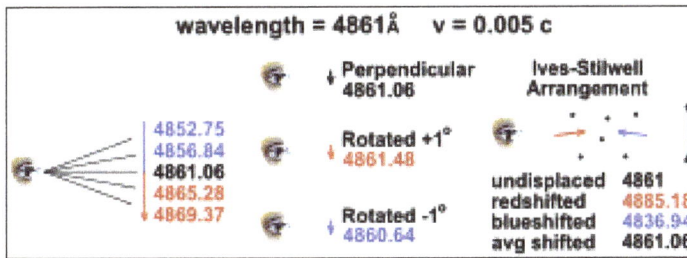

Why it is difficult to measure the transverse Doppler effect accurately using a transverse beam.

Einstein had initially suggested that the TDE might be measured by observing a beam of "canal rays" at right angles to the beam. Attempts to measure TDE following this scheme proved it to be impractical, since the maximum speed of particle beam available at the time was only a few thousandths of the speed of light.

Figure shows the results of attempting to measure the 4861 Angstrom line emitted by a beam of canal rays (a mixture of H1+, H2+, and H3+ ions) as they recombine with electrons stripped from the dilute hydrogen gas used to fill the Canal ray tube. Here, the predicted result of the TDE is a 4861.06 Angstrom line. On the left, longitudinal Doppler shift results in broadening the emission line to such an extent that the TDE cannot be observed. The middle figures illustrate that even if one narrows one's view to the exact center of the beam, very small deviations of the beam from an exact right angle introduce shifts comparable to the predicted effect.

Rather than attempt direct measurement of the TDE, Ives and Stilwell used a concave mirror that allowed them to simultaneously observe a nearly longitudinal direct beam (blue) and its reflected image (red). Spectroscopically, three lines would be observed: An undisplaced emission line, and blueshifted and redshifted lines. The average of the redshifted and blueshifted lines would be compared with the wavelength of the undisplaced emission line. The difference that Ives and Stilwell measured corresponded, within experimental limits, to the effect predicted by special relativity.

Various of the subsequent repetitions of the Ives and Stilwell experiment have adopted other strategies for measuring the mean of blueshifted and redshifted particle beam emissions. In some recent repetitions of the experiment, modern accelerator technology has been used to arrange for the observation of two counter-rotating particle beams. In other repetitions, the energies of gamma rays emitted by a rapidly moving particle beam have been measured at opposite angles relative to the direction of the particle beam. Since these experiments do not actually measure the wavelength of the particle beam at right angles to the beam, some authors have preferred to refer to the effect they are measuring as the "quadratic Doppler shift" rather than TDE.

Direct Measurement of Transverse Doppler Effect

The advent of particle accelerator technology has made possible the production of particle beams of considerably higher energy than was available to Ives and Stilwell. This has enabled the design of tests of the transverse Doppler effect directly along the lines of how Einstein originally envisioned them, i.e. by directly viewing a particle beam at a 90° angle. For example, Hasselkamp et al. observed the $H\alpha$ line emitted by hydrogen atoms moving at speeds ranging from 2.53×10^8 cm/s to 9.28×10^8 cm/s, finding the coefficient of the second order term in the relativistic approximation to be 0.52 ± 0.03, in excellent agreement with the theoretical value of $1/2$.

Other direct tests of the TDE on rotating platforms were made possible by the discovery of the Mössbauer effect, which enables the production of exceedingly narrow resonance lines for nuclear gamma ray emission and absorption. Mössbauer effect experiments have proven themselves easily capable of detecting TDE using emitter-absorber relative velocities on the order of 2×10^4 cm/s. These experiments include ones performed by Hay *et al.*, Champeney *et al.*, and Kündig.

Time Dilation Measurements

The transverse Doppler effect and the kinematic time dilation of special relativity are closely related. All validations of TDE represent validations of kinematic time dilation, and most validations of kinematic time dilation have also represented validations of TDE. An online resource, "What is the experimental basis of Special Relativity?" has documented, with brief commentary, many of the tests that, over the years, have been used to validate various aspects of special relativity. Kaivola et al. and McGowan et al. are examples of experiments classified in this resource as time dilation experiments. These two also represent tests of TDE. These experiments compared the frequency of two lasers, one locked to the frequency of a neon atom transition in a fast beam, the other locked to the same transition in thermal neon. The 1993 version of the experiment verified time dilation, and hence TDE, to an accuracy of 2.3×10^{-6}.

Relativistic Doppler Effect for Sound and Light

First-year physics textbooks almost invariably analyze Doppler shift for sound in terms of Newtonian kinematics, while analyzing Doppler shift for light and electromagnetic phenomena in terms of relativistic kinematics. This gives the false impression that acoustic phenomena requires a different analysis than light and radio waves.

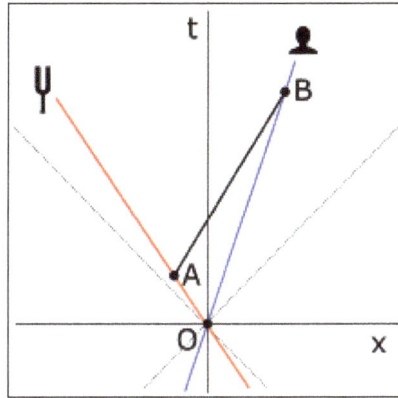

The relativistic Doppler shift formula is applicable to both sound and light.

The traditional analysis of the Doppler effect for sound represents a low speed approximation to the exact, relativistic analysis. The fully relativistic analysis for sound is, in fact, equally applicable to both sound and electromagnetic phenomena.

Consider the spacetime diagram in figure. Worldlines for a tuning fork (the source) and a receiver are both illustrated on this diagram. Events O and A represent two vibrations of the tuning fork. The period of the fork is the magnitude of OA, and the inverse slope of AB represents the speed of signal propagation (i.e. the speed of sound) to event B. We can therefore write:

$$c_s = \frac{x_B - x_A}{t_B - t_A} \text{ (speed of sound)}$$

$$v_s = -\frac{x_A}{t_A} \qquad v_r = \frac{x_B}{t_B} \text{ (speeds of source and receiver):}$$

$$|OA| = \sqrt{t_A^2 - (x_A/c)^2}$$

$$|OB| = \sqrt{t_B^2 - (x_B/c)^2}$$

v_s and v_r are assumed to be less than c_s since otherwise their passage through the medium will set up shock waves, invalidating the calculation. Some routine algebra gives the ratio of frequencies:

$$\frac{f_r}{f_s} = \frac{|OA|}{|OB|} = \frac{1 - v_r/c_s}{1 + v_s/c_s} \sqrt{\frac{1 - (v_s/c)^2}{1 - (v_r/c)^2}}$$

If v_r and v_s are small compared with c, the above equation reduces to the classical Doppler formula for sound.

If the speed of signal propagation c_s approaches c, it can be shown that the absolute speeds v_s and v_r of the source and receiver merge into a single relative speed independent of any reference to a fixed medium.

Indeed, we obtain $f_r = f_{r,s}\gamma = \frac{1 - \beta}{\sqrt{1 - \beta^2}} f_s = \sqrt{\frac{1 - \beta}{1 + \beta}} f_s$, the formula for relativistic longitudinal Doppler shift.

Analysis of the spacetime diagram in figure. gave a general formula for source and receiver moving directly along their line of sight, i.e. in collinear motion.

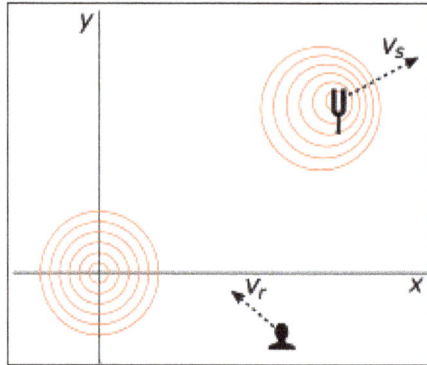

A source and receiver are moving in different directions and speeds in a frame
where the speed of sound is independent of direction.

Figure illustrates a scenario in two dimensions. The source moves with velocity v_s (at the time of emission). It emits a signal which travels at velocity C towards the receiver, which is traveling at velocity v_r at the time of reception. The analysis is performed in a coordinate system in which the signal's speed $|C|$ is independent of direction.

The ratio between the proper frequencies for the source and receiver is:

$$\frac{f_r}{f_s} = \frac{1 - \frac{|\mathbf{v_r}|}{|\mathbf{C}|}\cos(\theta_{C,v_r})}{1 - \frac{|\mathbf{v_s}|}{|\mathbf{C}|}\cos(\theta_{C,v_s})}\sqrt{\frac{1-(v_s/c)^2}{1-(v_r/c)^2}}$$

The leading ratio has the form of the classical Doppler effect, while the square root term represents the relativistic correction. If we consider the angles relative to the frame of the source, then $v_s = 0$ and the equation reduces to $f_r = \gamma(1 - \beta\cos\theta_s)f_s$, Einstein's 1905 formula for the Doppler effect. If we consider the angles relative to the frame of the receiver, then $v_r = 0$ and the equation reduces to $f_r = \dfrac{f_s}{\gamma(1+\beta\cos\theta_r)}$, the alternative form of the Doppler shift equation.

Thomas Precession

In physics, the Thomas precession, named after Llewellyn Thomas, is a relativistic correction that applies to the spin of an elementary particle or the rotation of a macroscopic gyroscope and relates the angular velocity of the spin of a particle following a curvilinear orbit to the angular velocity of the orbital motion.

For a given inertial frame, if a second frame is Lorentz-boosted relative to it, and a third boosted relative to the second, but non-colinear with the first boost, then the Lorentz transformation between the first and third frames involves a combined boost and rotation, known as the "Wigner rotation" or "Thomas rotation". For accelerated motion, the accelerated frame has an inertial frame at every instant. Two boosts a small time interval (as measured in the lab frame) apart leads to a Wigner rotation after the second boost. In the limit the time interval tends to zero, the accelerated frame will rotate at every instant, so the accelerated frame rotates with an angular velocity.

The precession can be understood geometrically as a consequence of the fact that the space of velocities in relativity is hyperbolic, and so parallel transport of a vector (the gyroscope's angular velocity) around a circle (its linear velocity) leaves it pointing in a different direction, or understood algebraically as being a result of the non-commutativity of Lorentz transformations. Thomas precession gives a correction to the spin–orbit interaction in quantum mechanics, which takes into account the relativistic time dilation between the electron and the nucleus of an atom.

Thomas precession is a kinematic effect in the flat spacetime of special relativity. In the curved spacetime of general relativity, Thomas precession combines with a geometric effect to produce de Sitter precession. Although Thomas precession (*net rotation after a trajectory that returns to its initial velocity*) is a purely kinematic effect, it only occurs in curvilinear motion and therefore cannot be observed independently of some external force causing the curvilinear motion such as that caused by an electromagnetic field, a gravitational field or a mechanical force, so Thomas precession is usually accompanied by dynamical effects.

If the system experiences no external torque, e.g., in external scalar fields, its spin dynamics is determined only by the Thomas precession. A single discrete Thomas rotation (as opposed to the series of infinitesimal rotations that add up to the Thomas precession) is present in situations anytime there are three or more inertial frames in non-collinear motion, as can be seen using Lorentz transformations.

Consider a physical system moving through Minkowski spacetime. Assume that there is at any moment an inertial system such that in it, the system is at rest. This assumption is sometimes called the third postulate of relativity. This means that at any instant, the coordinates and state of the system can be Lorentz transformed to the lab system through *some* Lorentz transformation.

Let the system be subject to *external forces* that produce no torque with respect to its center of mass in its (instantaneous) rest frame. The condition of "no torque" is necessary to isolate the phenomenon of Thomas precession. As a simplifying assumption one assumes that the external forces bring the system back to its initial velocity after some finite time. Fix a Lorentz frame O such that the initial and final velocities are zero.

The Pauli–Lubanski spin vector S_μ is defined to be $(0, S_i)$ in the system's *rest* frame, with S_i the angular-momentum three-vector about the center of mass. In the motion from initial to final position, S_μ undergoes a rotation, as recorded in O, from its initial to its final value. This continuous change is the Thomas precession.

Statement

Consider the motion of a particle. Introduce a lab frame Σ in which an observer can measure the relative motion of the particle. At each instant of time the particle has an inertial frame in which it is at rest. Relative to this lab frame, the instantaneous velocity of the particle is $v(t)$ with magnitude $|v| = v$ bounded by the speed of light c, so that $0 \le v < c$. Here the time t is the coordinate time as measured in the lab frame, *not* the proper time of the particle.

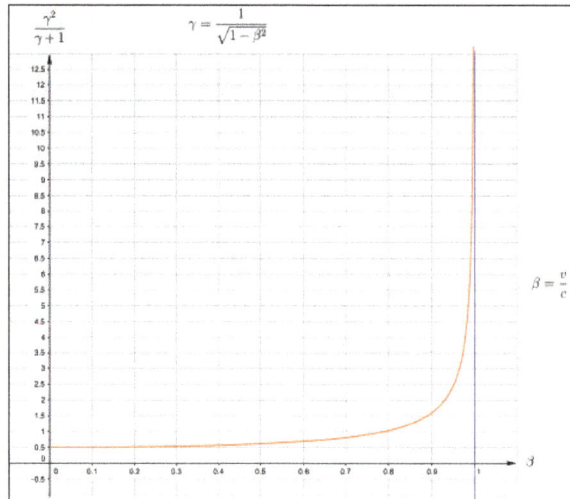

Value of $\gamma^2/(\gamma + 1)$ as $\beta = v/c$ increases, with v the instantaneous magnitude of the particle's velocity. The Thomas rotation is negligible for $\beta < 0.5$, increases steadily for $0.5 < \beta < 0.8$, then rapidly shoots to infinity as β tends to 1. The "Thomas half" is evident in the low-speed limit, and the rotation is only very clear for speeds approaching that of light.

Apart from the upper limit on magnitude, the velocity of the particle is arbitrary and not necessarily constant, its corresponding vector of acceleration is a = $dv(t)/dt$. As a result of the Wigner rotation at every instant, the particle's frame precesses with an angular velocity given by the:

Thomas Precession:

$$\omega_{\text{T}} = \frac{1}{c^2}\left(\frac{\gamma^2}{\gamma+1}\right)\mathbf{a}\times\mathbf{v}$$

where × is the cross product and:

$$\gamma = \frac{1}{\sqrt{1-\dfrac{|\mathbf{v}(t)|^2}{c^2}}}$$

is the instantaneous Lorentz factor, a function of the particle's instantaneous velocity. Like any angular velocity, ω_{T} is a pseudovector; its magnitude is the angular speed the particle's frame precesses (in radians per second), and the direction points along the rotation axis. As is usual, the right-hand convention of the cross product is used.

The precession depends on *accelerated* motion, and the non-collinearity of the particle's instantaneous velocity and acceleration. No precession occurs if the particle moves with uniform velocity (constant v so a = 0), or accelerates in a straight line (in which case v and a are parallel or antiparallel so their cross product is zero). The particle has to move in a curve, say an arc, spiral, helix, or a circular orbit or elliptical orbit, for its frame to precess. The angular velocity of the precession is a maximum if the velocity and acceleration vectors are perpendicular throughout the motion (a circular orbit), and is large if their magnitudes are large (the magnitude of v is almost c).

In the non-relativistic limit, $v \to 0$ so $\gamma \to 1$, and the angular velocity is approximately:

$$\omega_T \approx \frac{1}{2c^2}\mathbf{a}\times\mathbf{v}$$

The factor of 1/2 turns out to be the critical factor to agree with experimental results. It is informally known as the "Thomas half".

Lorentz Transformations

The description of relative motion involves Lorentz transformations, and it is convenient to use them in matrix form; symbolic matrix expressions summarize the transformations and are easy to manipulate, and when required the full matrices can be written explicitly. Also, to prevent extra factors of c cluttering the equations, it is convenient to use the definition $\beta(t) = v(t)/c$ with magnitude $|\beta| = \beta$ such that $0 \leq \beta < 1$.

The spacetime coordinates of the lab frame are collected into a 4×1 column vector, and the boost is represented as a 4×4 symmetric matrix, respectively,

$$X = \begin{bmatrix} ct \\ x \\ y \\ z \end{bmatrix}, \quad B(\beta) = \begin{bmatrix} \gamma & -\gamma\beta_x & -\gamma\beta_y & -\gamma\beta_z \\ -\gamma\beta_x & 1+(\gamma-1)\frac{\beta_x^2}{\beta^2} & (\gamma-1)\frac{\beta_x\beta_y}{\beta^2} & (\gamma-1)\frac{\beta_x\beta_z}{\beta^2} \\ -\gamma\beta_y & (\gamma-1)\frac{\beta_y\beta_x}{\beta^2} & 1+(\gamma-1)\frac{\beta_y^2}{\beta^2} & (\gamma-1)\frac{\beta_y\beta_z}{\beta^2} \\ -\gamma\beta_z & (\gamma-1)\frac{\beta_z\beta_x}{\beta^2} & (\gamma-1)\frac{\beta_z\beta_y}{\beta^2} & 1+(\gamma-1)\frac{\beta_z^2}{\beta^2} \end{bmatrix}$$

and turn:

$$\gamma = \frac{1}{\sqrt{1-|\beta|^2}}$$

is the Lorentz factor of β. In other frames, the corresponding coordinates are also arranged into column vectors. The inverse matrix of the boost corresponds to a boost in the opposite direction, and is given by $B(\beta)^{-1} = B(-\beta)$.

At an instant of lab-recorded time t measured in the lab frame, the transformation of spacetime coordinates from the lab frame Σ to the particle's frame Σ' is:

$$X' = B(\beta)X$$

and at later lab-recorded time $t + \Delta t$ we can define a new frame Σ'' for the particle, which moves with velocity $\beta + \Delta\beta$ relative to Σ, and the corresponding boost is:

$$X'' = B(\beta+\Delta\beta)X$$

The vectors β and $\Delta\beta$ are two separate vectors. The latter is a small increment, and can be conveniently split into components parallel ($\|$) and perpendicular ($\perp$) to β:

$$\Delta\beta = \Delta\beta_{\|} + \Delta\beta_{\perp}$$

Combining $X' = B(\beta)X$ and $X'' = B(\beta + \Delta\beta)X$ obtains the Lorentz transformation between Σ' and Σ'',

$$X'' = B(\beta + \Delta\beta)B(-\beta)X',$$

and this composition contains all the required information about the motion between these two lab times. Notice $B(\beta + \Delta\beta)B(-\beta)$ and $B(\beta + \Delta\beta)$ are infinitesimal transformations because they involve a small increment in the relative velocity, while $B(-\beta)$ is not.

The composition of two boosts equates to a single boost combined with a Wigner rotation about an axis perpendicular to the relative velocities;

$$\Lambda = B(\beta + \Delta\beta)B(-\beta) = R(\Delta\theta)B(\Delta\mathbf{b})$$

The rotation is given by is a 4×4 rotation matrix R in the axis–angle representation, and coordinate systems are taken to be right-handed. This matrix rotates 3d vectors anticlockwise about an axis (active transformation), or equivalently rotates coordinate frames clockwise about the same axis (passive transformation). The axis-angle vector $\Delta\theta$ parametrizes the rotation, its magnitude $\Delta\theta$ is the angle Σ'' has rotated, and direction is parallel to the rotation axis, in this case the axis is parallel to the cross product $(-\beta) \times (\beta + \Delta\beta) = -\beta \times \Delta\beta$. If the angles are negative, then the sense of rotation is reversed. The inverse matrix is given by $R(\Delta\theta)^{-1} = R(-\Delta\theta)$.

Corresponding to the boost is the (small change in the) boost vector Δb, with magnitude and direction of the relative velocity of the boost (divided by c). The boost $B(\Delta b)$ and rotation $R(\Delta\theta)$ here are infinitesimal transformations because Δb and rotation $\Delta\theta$ are small.

The rotation gives rise to the Thomas precession, but there is a subtlety. To interpret the particle's frame as a co-moving inertial frame relative to the lab frame, and agree with the non-relativistic limit, we expect the transformation between the particle's instantaneous frames at times t and $t + \Delta t$ to be related by a boost *without* rotation. Combining above equations and rearranging gives:

$$B(\Delta\mathbf{b})X' = R(-\Delta\theta)X'' = X''',$$

where another instantaneous frame Σ''' is introduced with coordinates X''', to prevent conflation with Σ''. To summarize the frames of reference: in the lab frame Σ an observer measures the motion of the particle, and three instantaneous inertial frames in which the particle is at rest are Σ' (at time t), Σ'' (at time $t + \Delta t$), and Σ''' (at time $t + \Delta t$). The frames Σ'' and Σ''' are at the same location and time, they differ only by a rotation. By contrast Σ' and Σ''' differ by a boost and lab time interval Δt.

Relating the coordinates X''' to the lab coordinates X:

$$X''' = R(-\Delta\theta)X'' = R(-\Delta\theta)B(\beta + \Delta\beta)X,$$

The frame Σ''' is rotated in the negative sense.

The rotation is between two instants of lab time. As $\Delta t \to 0$, the particle's frame rotates at every instant, and the continuous motion of the particle amounts to a continuous rotation with an angular velocity at every instant. Dividing $-\Delta\theta$ by Δt, and taking the limit $\Delta t \to 0$, the angular velocity is by definition:

$$\omega_T = -\lim_{\Delta t \to 0} \frac{\Delta\theta}{\Delta t}$$

It remains to find what $\Delta\theta$ precisely is.

Extracting the Formula

The composition can be obtained by explicitly calculating the matrix product. The boost matrix of $\beta + \Delta\beta$ will require the magnitude and Lorentz factor of this vector. Since $\Delta\beta$ is small, terms of "second order" $|\Delta\beta|^2, (\Delta\beta_x)^2, (\Delta\beta_y)^2, \Delta\beta_x\Delta\beta_y$ and higher are negligible. Taking advantage of this fact, the magnitude squared of the vector is:

$$|\beta + \Delta\beta|^2 = |\beta|^2 + 2\beta\cdot\Delta\beta$$

and expanding the Lorentz factor of $\beta + \Delta\beta$ as a power series gives to first order in $\Delta\beta$:

$$\frac{1}{\sqrt{1-|\beta+\Delta\beta|^2}} = 1 + \frac{1}{2}|\beta+\Delta\beta|^2 + \frac{3}{8}|\beta+\Delta\beta|^4 + \cdots$$

$$= \left(1 + \frac{|\beta|^2}{2} + \frac{3}{8}|\beta|^4 + \cdots\right) + \left(1 + \frac{3}{2}|\beta|^2 + \cdots\right)\beta\cdot\Delta\beta$$

$$\approx \gamma + \gamma^3\beta\cdot\Delta\beta$$

using the Lorentz factor γ of β as above.

Composition of Boosts in the XY Plane

To simplify the calculation without loss of generality, take the direction of β to be entirely in the x direction, and $\Delta\beta$ in the xy plane, so the parallel component is along the x direction while the perpendicular component is along the y direction. The axis of the Wigner rotation is along the z direction. In the Cartesian basis e_x, e_y, e_z, a set of mutually perpendicular unit vectors in their indicated directions, we have:

$$\beta = \beta e_x, \quad \Delta\beta_\parallel = \Delta\beta_x e_x, \quad \Delta\beta_\perp = \Delta\beta_y e_y, \quad \beta\times\Delta\beta = \beta\Delta\beta_y e_z$$

This simplified setup allows the boost matrices to be given explicitly with the minimum number of matrix entries. In general, of course, β and $\Delta\beta$ can be in any plane, the final result given later will not be different.

Explicitly, at time t the boost is in the negative x direction:

$$B(-\beta) = \begin{bmatrix} \gamma & \gamma\beta & 0 & 0 \\ \gamma\beta & \gamma & 0 & 0 \\ 0 & 0 & 1 & 0 \\ 0 & 0 & 0 & 1 \end{bmatrix}$$

and the boost at the time $t + \Delta t$ is:

$$B(\beta + \Delta\beta) = \begin{bmatrix} \gamma + \gamma^3\beta\Delta\beta_x & -(\gamma\beta + \gamma^3\Delta\beta_x) & -\gamma\Delta\beta_y & 0 \\ -(\gamma\beta + \gamma^3\Delta\beta_x) & \gamma + \gamma^3\beta\Delta\beta_x & \left(\dfrac{\gamma-1}{\beta}\right)\Delta\beta_y & 0 \\ -\gamma\Delta\beta_y & \left(\dfrac{\gamma-1}{\beta}\right)\Delta\beta_y & 1 & 0 \\ 0 & 0 & 0 & 1 \end{bmatrix}$$

where γ is the Lorentz factor of β, *not* $\beta + \Delta\beta$. The composite transformation is then the matrix product:

$$\Lambda = B(\beta + \Delta\beta)B(-\beta) = \begin{bmatrix} 1 & -\gamma^2\Delta\beta_x & -\gamma\Delta\beta_y & 0 \\ -\gamma^2\Delta\beta_x & 1 & \left(\dfrac{\gamma-1}{\beta}\right)\Delta\beta_y & 0 \\ -\gamma\Delta\beta_y & -\left(\dfrac{\gamma-1}{\beta}\right)\Delta\beta_y & 1 & 0 \\ 0 & 0 & 0 & 1 \end{bmatrix}$$

Introducing the boost generators:

$$K_x = \begin{bmatrix} 0 & 1 & 0 & 0 \\ 1 & 0 & 0 & 0 \\ 0 & 0 & 0 & 0 \\ 0 & 0 & 0 & 0 \end{bmatrix}, \quad K_y = \begin{bmatrix} 0 & 0 & 1 & 0 \\ 0 & 0 & 0 & 0 \\ 1 & 0 & 0 & 0 \\ 0 & 0 & 0 & 0 \end{bmatrix}, \quad K_z = \begin{bmatrix} 0 & 0 & 0 & 1 \\ 0 & 0 & 0 & 0 \\ 0 & 0 & 0 & 0 \\ 1 & 0 & 0 & 0 \end{bmatrix}$$

and rotation generators:

$$J_x = \begin{bmatrix} 0 & 0 & 0 & 0 \\ 0 & 0 & 0 & 0 \\ 0 & 0 & 0 & -1 \\ 0 & 0 & 1 & 0 \end{bmatrix}, \quad J_y = \begin{bmatrix} 0 & 0 & 0 & 0 \\ 0 & 0 & 0 & 1 \\ 0 & 0 & 0 & 0 \\ 0 & -1 & 0 & 0 \end{bmatrix}, \quad J_z = \begin{bmatrix} 0 & 0 & 0 & 0 \\ 0 & 0 & -1 & 0 \\ 0 & 1 & 0 & 0 \\ 0 & 0 & 0 & 0 \end{bmatrix}$$

along with the dot product · facilitates the coordinate independent expression:

$$\Lambda = I - \left(\frac{\gamma-1}{\beta^2}\right)(\beta \times \Delta\beta)\cdot \mathbf{J} - \gamma(\gamma\Delta\beta_\parallel + \Delta\beta_\perp)\cdot \mathbf{K}$$

which holds if β and $\Delta\beta$ lie in any plane. This is an infinitesimal Lorentz transformation in the form of a combined boost and rotation:

$$\Lambda = I - \Delta\theta \cdot \mathbf{J} - \Delta\mathbf{b} \cdot \mathbf{K}$$

where,

$$\Delta\theta = \left(\frac{\gamma-1}{\beta^2}\right)\beta \times \Delta\beta = \frac{1}{c^2}\left(\frac{\gamma^2}{\gamma+1}\right)\mathbf{v} \times \Delta\mathbf{v}$$

$$\Delta\mathbf{b} = \gamma(\gamma\Delta\beta_\parallel + \Delta\beta_\perp)$$

After dividing $\Delta\theta$ by Δt, one obtains the instantaneous angular velocity:

$$\omega_T = \frac{1}{c^2}\left(\frac{\gamma^2}{\gamma+1}\right)\mathbf{a} \times \mathbf{v}$$

where a is the acceleration of the particle as observed in the lab frame. No forces were specified or used in the derivation so the precession is a kinematical effect - it arises from the geometric aspects of motion. However, forces cause accelerations, so the Thomas precession is observed if the particle is subject to forces.

Thomas precession can also be derived using the Fermi-Walker transport equation. One assumes uniform circular motion in flat Minkowski spacetime. The spin 4-vector is orthogonal to the velocity 4-vector. Fermi-Walker transport preserves this relation. One finds that the dot product of the acceleration 4-vector with the spin 4-vector varies sinusoidally with time with an angular frequency 'Y ω, where ω is the angular frequency of the circular motion and 'Y=1/√⟨1-v^2/c^2⟩. This is easily shown by taking the second time derivative of that dot product. Because this angular frequency exceeds ω, the spin precesses in the retrograde direction. The difference $(\gamma-1)\omega$ is the Thomas precession angular frequency already given, as is simply shown by realizing that that the magnitude of the 3-acceleration is ω v.

Applications

In Electron Orbitals

In quantum mechanics Thomas precession is a correction to the spin-orbit interaction, which takes into account the relativistic time dilation between the electron and the nucleus in hydrogenic atoms.

Basically, it states that spinning objects precess when they accelerate in special relativity because

Lorentz boosts do not commute with each other.

To calculate the spin of a particle in a magnetic field, one must also take into account Larmor precession.

In a Foucault Pendulum

The rotation of the swing plane of Foucault pendulum can be treated as a result of parallel transport of the pendulum in a 2-dimensional sphere of Euclidean space. The hyperbolic space of velocities in Minkowski spacetime represents a 3-dimensional (pseudo-) sphere with imaginary radius and imaginary timelike coordinate. Parallel transport of a spinning particle in relativistic velocity space leads to Thomas precession, which is similar to the rotation of the swing plane of a Foucault pendulum. The angle of rotation in both cases is determined by the area integral of curvature in agreement with the Gauss–Bonnet theorem.

Thomas precession gives a correction to the precession of a Foucault pendulum. For a Foucault pendulum located in the city of Nijmegen in the Netherlands the correction is:

$$\omega \approx 9.5 \cdot 10^{-7} \text{ arcseconds / day.}$$

Note that it is more than two orders of magnitude smaller than the precession due to the general-relativistic correction arising from frame-dragging, the Lense–Thirring precession.

Ladder Paradox

The ladder paradox (or barn-pole paradox) is a thought experiment in special relativity. It involves a ladder, parallel to the ground, travelling horizontally at relativistic speed (near the speed of light) and therefore undergoing a Lorentz length contraction. The ladder is imagined passing through the open front and rear doors of a garage or barn which is shorter than its rest length, so if the ladder was not moving it would not be able to fit inside. To a stationary observer, due to the contraction, the moving ladder is able to fit entirely inside the building as it passes through. On the other hand, from the point of view of an observer moving with the ladder, the ladder will not be contracted, and it is the building which will be Lorentz contracted to an even smaller length. Therefore the ladder will not be able to fit inside the building as it passes through. This poses an apparent discrepancy between the realities of both observers.

This apparent paradox results from the mistaken assumption of absolute simultaneity. The ladder is said to fit into the garage if both of its ends can be made to be simultaneously inside the garage. The paradox is resolved when it is considered that in relativity, simultaneity is relative to each observer, making the answer to whether the ladder fits inside the garage also relative to each of them.

Paradox

The simplest version of the problem involves a garage, with a front and back door which are open, and a ladder which, when at rest with respect to the garage, is too long to fit inside. We now move the ladder at a high horizontal velocity through the stationary garage. Because of its high velocity, the ladder undergoes the relativistic effect of length contraction, and becomes significantly shorter. As a result, as the ladder passes through the garage, it is, for a time, completely contained inside it. We could, if we liked, simultaneously close both doors for a brief time, to demonstrate that the ladder fits.

So far, this is consistent. The apparent paradox comes when we consider the symmetry of the situation. As an observer moving with the ladder is travelling at constant velocity in the inertial reference frame of the garage, this observer also occupies an inertial frame, where, by the principle of relativity, the same laws of physics apply. From this perspective, it is the ladder which is now stationary, and the garage which is moving with high velocity. It is therefore the garage which is length contracted, and we now conclude that it is far too small to have ever fully contained the ladder as it passed through: the ladder does not fit, and we cannot close both doors on either side of the ladder without hitting it. This apparent contradiction is the paradox.

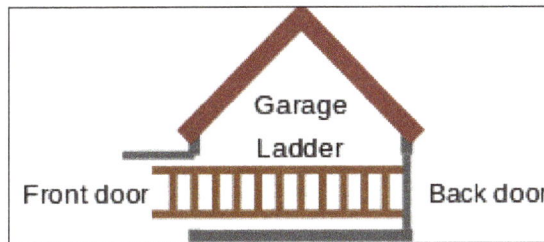

An overview of the garage and the ladder at rest.

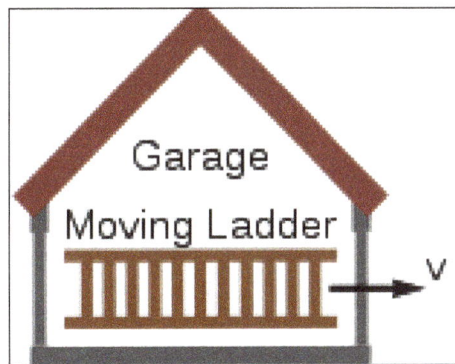

In the garage frame, the ladder undergoes length contraction and
will therefore fit into the garage.

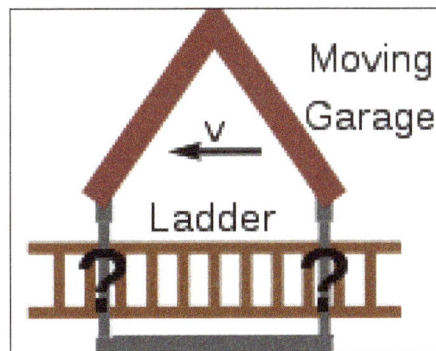

In the ladder frame, the garage undergoes length contraction and is
too small to contain the ladder.

Resolution

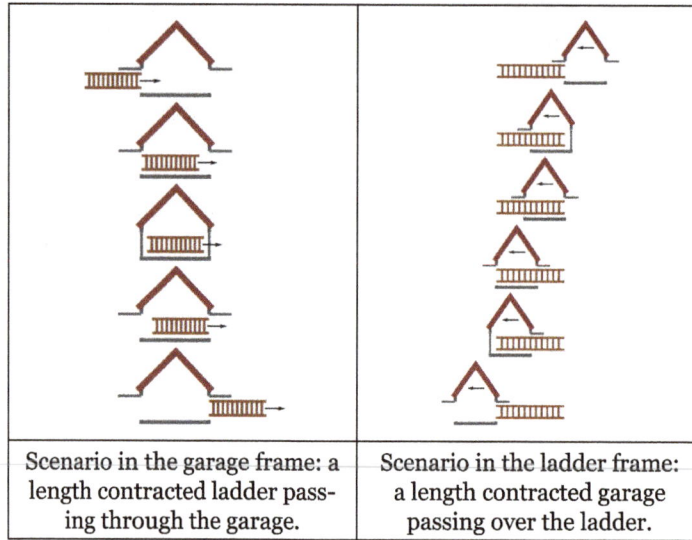

| Scenario in the garage frame: a length contracted ladder passing through the garage. | Scenario in the ladder frame: a length contracted garage passing over the ladder. |

The solution to the apparent paradox lies in the relativity of simultaneity: what one observer (e.g. with the garage) considers to be two simultaneous events may not in fact be simultaneous to another observer (e.g. with the ladder). When we say the ladder "fits" inside the garage, what we mean precisely is that, at some specific time, the position of the back of the ladder and the position of the front of the ladder were both inside the garage; in other words, the front and back of the ladder were inside the garage simultaneously. As simultaneity is relative, then, two observers disagree on whether the ladder fits. To the observer with the garage, the back end of the ladder was in the garage at the same time that the front end of the ladder was, and so the ladder fit; but to the observer with the ladder, these two events were not simultaneous, and the ladder did not fit.

A clear way of seeing this is to consider the doors, which, in the frame of the garage, close for the brief period that the ladder is fully inside. We now look at these events in the frame of the ladder. The first event is the front of the ladder approaching the exit door of the garage. The door closes, and then opens again to let the front of the ladder pass through. At a later time, the back of the ladder passes through the entrance door, which closes and then opens. We see that, as simultaneity is relative, the two doors did not need to be shut at the same time, and the ladder did not need to fit inside the garage.

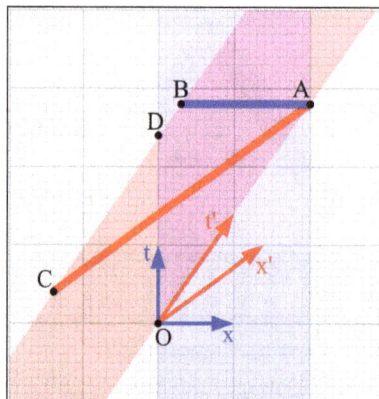

A Minkowski diagram of ladder paradox.

The situation can be further illustrated by the Minkowski diagram. The diagram is in the rest frame of the garage. The vertical light-blue band shows the garage in space-time, and the light-red band shows the ladder in space-time. The x and t axes are the garage space and time axes, respectively, and x' and t' are the ladder space and time axes, respectively.

The garage is shown in light blue, the ladder in light red. The diagram is in the rest frame of the garage, with x and t being the garage space and time axes, respectively. The ladder frame is for a person sitting on the front of the ladder, with x' and t' being the ladder space and time axes respectively. The blue and red lines, AB and AC, depict the ladder at the time when its front end meets the garage's exit door, in the frame of reference of the garage and the ladder, respectively. Event D is the rear end of the ladder reaching the garage's entrance.

In the frame of the garage, the ladder at any specific time is represented by a horizontal set of points, parallel to the x axis, in the red band. One example is the bold blue line segment, which lies inside the blue band representing the garage, and which represents the ladder at a time when it is fully inside the garage. In the frame of the ladder, however, sets of simultaneous events lie on lines parallel to the x' axis; the ladder at any specific time is therefore represented by a cross section of such a line with the red band. One such example is the bold red line segment. We see that such line segments never lie fully inside the blue band; that is, the ladder never lies fully inside the garage.

Shutting the Ladder in the Garage

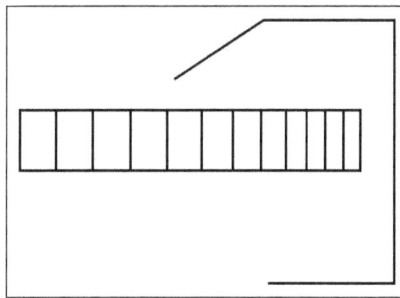

A ladder contracting under acceleration to fit into a length contracted garage.

In a more complicated version of the paradox, we can physically trap the ladder once it is fully inside the garage. This could be done, for instance, by not opening the exit door again after we close it. In the frame of the garage, we assume the exit door is immovable, and so when the ladder hits it, we say that it instantaneously stops. By this time, the entrance door has also closed, and so the ladder is stuck inside the garage. As its relative velocity is now zero, it is not length contracted, and is now longer than the garage; it will have to bend, snap, or explode.

Again, the puzzle comes from considering the situation from the frame of the ladder. In the above analysis, in its own frame, the ladder was always longer than the garage. So how did we ever close the doors and trap it inside?

It is worth noting here a general feature of relativity: we have deduced, by considering the frame of the garage, that we do indeed trap the ladder inside the garage. This must therefore be true in any frame - it cannot be the case that the ladder snaps in one frame but not in another. From the ladder's frame, then, we know that there must be some explanation for how the ladder came to be trapped; we must simply find the explanation.

The explanation is that, although all parts of the ladder simultaneously decelerate to zero in the garage's frame, because simultaneity is relative, the corresponding decelerations in the frame of the ladder are not simultaneous. Instead, each part of the ladder decelerates sequentially, from front to back, until finally the back of the ladder decelerates, by which time it is already within the garage.

As length contraction and time dilation are both controlled by the Lorentz transformations, the ladder paradox can be seen as a physical correlate of the twin paradox, in which instance one of a set of twins leaves earth, travels at speed for a period, and returns to earth a bit younger than the earthbound twin. As in the case of the ladder trapped inside the barn, if neither frame of reference is privileged — each is moving only relative to the other — how can it be that it's the traveling twin and not the stationary one who is younger (just as it's the ladder rather than the barn which is shorter)? In both instances it is the acceleration-deceleration that differentiates the phenomena: it's the twin, not the earth (or the ladder, not the barn) that undergoes the force of deceleration in returning to the temporal (or physical, in the case of the ladder-barn) inertial frame.

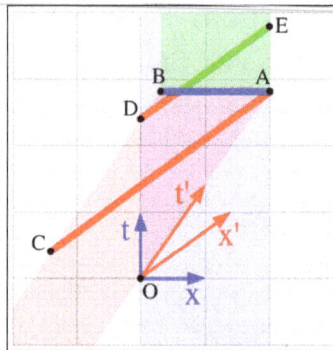

A Minkowski diagram of the case where the ladder is stopped all along its length, simultaneously in the garage frame. When this occurs, the garage frame sees the ladder as AB, but the ladder frame sees the ladder as AC. When the back of the ladder enters the garage at point D, it has not yet felt the effects of the acceleration of its front end. At this time, according to someone at rest with respect to the back of the ladder, the front of the ladder will be at point E and will see the ladder as DE. It is seen that this length in the ladder frame is not the same as CA, the rest length of the ladder before the deceleration.

Ladder Paradox and Transmission of Force

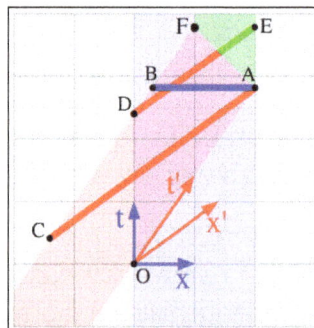

A Minkowski diagram of the case where the ladder is stopped by impact with the back wall of the garage. The impact is event A. At impact, the garage frame sees the ladder as AB, but the ladder frame sees the ladder as AC. The ladder does not move out of the garage, so its front end now goes directly upward, through point E. The back of the ladder will not change its trajectory in space-time until it feels the effects of the impact. The effect of the impact can propagate outward from A no faster than the speed of light, so the back of the ladder will never feel the effects of the impact until point F or later, at which time the ladder is well within the garage in both frames. Note that when the diagram is drawn in the frame of the ladder, the speed of light is the same, but the ladder is longer, so it takes more time for the force to reach the back end; this gives enough time for the back of the ladder to move inside the garage.

What if the back door (the door the ladder exits out of) is closed permanently and does not open? Suppose that the door is so solid that the ladder will not penetrate it when it collides, so it must stop. Then, as in the scenario described above, in the frame of reference of the garage, there is a moment when the ladder is completely within the garage (i.e., the back of the ladder is inside the front door), before it collides with the back door and stops. However, from the frame of reference of the ladder, the ladder is too big to fit in the garage, so by the time it collides with the back door and stops, the back of the ladder still has not reached the front door. This seems to be a paradox. The question is, does the back of the ladder cross the front door or not?

The difficulty arises mostly from the assumption that the ladder is rigid (i.e., maintains the same shape). Ladders seem rigid in everyday life. But being completely rigid requires that it can transfer force at infinite speed (i.e., when you push one end the other end must react immediately, otherwise the ladder will deform). This contradicts special relativity, which states that information can travel no faster than the speed of light (which is too fast for us to notice in real life, but is significant in the ladder scenario). So objects cannot be perfectly rigid under special relativity.

In this case, by the time the front of the ladder collides with the back door, the back of the ladder does not know it yet, so it keeps moving forwards (and the ladder "compresses"). In both the frame of the garage and the inertial frame of the ladder, the back end keeps moving at the time of the collision, until at least the point where the back of the ladder comes into the light cone of the collision (i.e., a point where force moving backwards at the speed of light from the point of the collision will reach it). At this point the ladder is actually shorter than the original contracted length, so the back end is well inside the garage. Calculations in both frames of reference will show this to be the case.

What happens after the force reaches the back of the ladder is not specified. Depending on the physics, the ladder could break; or, if it were sufficiently elastic, it could bend and re-expand to its original length. At sufficiently high speeds, any realistic material would violently explode into a plasma.

Man Falling into Grate Variation

This paradox was originally proposed and solved by Wolfgang Rindler and involved a fast walking man, represented by a rod, falling into a grate. It is assumed that the rod is entirely over the grate in the grate frame of reference before the downward acceleration begins simultaneously and equally applied to each point in the rod.

From the perspective of the grate, the rod undergoes a length contraction and fits into the grate. However, from the perspective of the rod, it is the *grate* undergoing a length contraction, through which it seems the rod is then too long to fall.

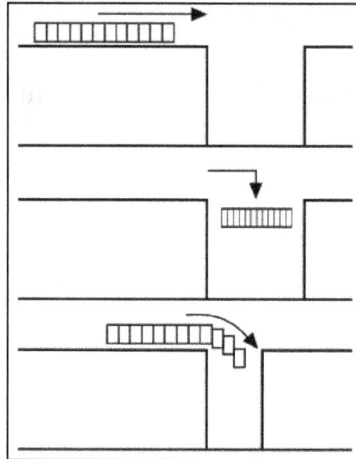

A man (represented by a segmented rod) falling into a grate.

The downward acceleration of the rod, which is simultaneous in the grate's frame of reference, is not simultaneous in the rod's frame of reference. In the rod's frame of reference, the front of the rod is first accelerated downward, and as time goes by, more and more of the rod is subjected to the downward acceleration, until finally the back of the rod is accelerated downward. This results in a bending of the rod in the rod's frame of reference. Since this bending occurs in the rod's rest frame, it is a true physical distortion of the rod which will cause stresses to occur in the rod.

For this non-rigid behaviour of the rod to become apparent, both the rod itself and the grate must be of such a scale that the traversal time is measurable.

Bar and Ring Paradox

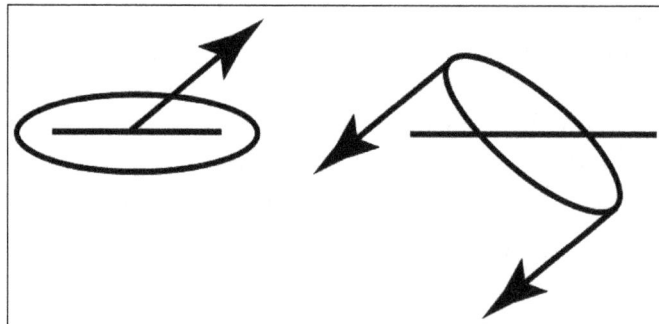

The diagram on the left illustrates a bar and a ring in the rest frame of the ring at the instant that their centers coincide. The bar is Lorentz-contracted and moving upward and to the right while the ring is stationary and uncontracted. The diagram on the right illustrates the situation at the same instant, but in the rest frame of the bar. The ring is now Lorentz-contracted and rotated with respect to the bar, and the bar is uncontracted. Again, the ring passes over the bar without touching it.

A problem very similar but simpler than the rod and grate paradox, involving only inertial frames, is the "bar and ring" paradox. The rod and grate paradox is complicated: it involves non-inertial frames of reference since at one moment the man is walking horizontally, and a moment later he is falling downward; and it involves a physical deformation of the man (or segmented rod), since the rod is bent in one frame of reference and straight in another. These aspects of the problem introduce complications involving the stiffness of the rod which tends to obscure the real nature of the "paradox". The "bar and ring" paradox is free of these complications: a bar, which is slightly larger in length than the diameter of a ring, is moving upward and to the right with its long axis horizontal, while the ring is stationary and the plane of the ring is also horizontal. If the motion of the bar is such that the center of the bar coincides with the center of the ring at some point in time, then the bar will be Lorentz-contracted due to the forward component of its motion, and it will pass through the ring. The paradox occurs when the problem is considered in the rest frame of the bar. The ring is now moving downward and to the left, and will be Lorentz-contracted along its horizontal length, while the bar will not be contracted at all. How can the bar pass through the ring?

The resolution of the paradox again lies in the relativity of simultaneity. The length of a physical object is defined as the distance between two *simultaneous* events occurring at each end of the body, and since simultaneity is relative, so is this length. This variability in length is just the Lorentz contraction. Similarly, a physical angle is defined as the angle formed by three *simultaneous* events, and this angle will also be a relative quantity. In the above paradox, although the rod and the plane of the ring are parallel in the rest frame of the ring, they are not parallel in the rest frame of the rod. The uncontracted rod passes through the Lorentz-contracted ring because the plane of the ring is rotated relative to the rod by an amount sufficient to let the rod pass through.

In mathematical terms, a Lorentz transformation can be separated into the product of a spatial rotation and a "proper" Lorentz transformation which involves no spatial rotation. The mathematical resolution of the bar and ring paradox is based on the fact that the product of two proper Lorentz transformations (horizontal and vertical) may produce a Lorentz transformation which is not proper (diagonal) but rather includes a spatial rotation component.

Twin Paradox

The twin paradox is a thought experiment that demonstrates the curious manifestation of time dilation in modern physics, as it was introduced by Albert Einstein through the theory of relativity.

Consider two twins, named Biff and Cliff. On their 20th birthday, Biff decides to get in a spaceship and take off into outer space, traveling at nearly the speed of light. He journeys around the cosmos at this speed for about 5 years, returning to the Earth when he is 25 years old.

Cliff, on the other hand, remains on the Earth. When Biff returns, it turns out that Cliff is 95 years old.

According to relativity, two frames of reference that move differently from each other experience time differently, a process known as time dilation. Because Biff was moving so rapidly, time was in effect moving slower for him. This can be calculated precisely using Lorentz transformations, which are a standard part of relativity.

Twin Paradox One

The first twin paradox isn't really a scientific paradox, but a logical one: How old is Biff?

Biff has experienced 25 years of life, but he was also born the same moment as Cliff, which was 90 years ago. So is he 25 years old or 90 years old?

In this case, the answer is "both" depending on which way you're measuring age. According to his driver's license, which measures Earth time (and is no doubt expired), he's 90. According to his body, he's 25. Neither age is "right" or "wrong," although the social security administration might take exception if he tries to claim benefits.

Twin Paradox Two

The second paradox is a bit more technical, and really comes to the heart of what physicists mean when they talk about relativity. The entire scenario is based on the idea that Biff was traveling very fast, so time slowed down for him.

The problem is that in relativity, only the relative motion is involved. So what if you considered things from Biff's point of view, then he stayed stationary the whole time, and it was Cliff who was moving away at rapid speeds. Shouldn't calculations performed in this way mean that Cliff is the one who ages more slowly? Doesn't relativity imply that these situations are symmetrical?

Now, if Biff and Cliff were on spaceships traveling at constant speeds in opposite directions, this argument would be perfectly true. The rules of special relativity, which govern constant speed (inertial) frames of reference, indicate that only the relative motion between the two is what matters. In fact, if you're moving at a constant speed, there's not even an experiment that you can perform within your frame of reference which would distinguish you from being at rest. (Even if you looked outside the ship and compared yourself to some other constant frame of reference, you could only determine that *one of you* is moving, but not which one.)

But there's one very important distinction here: Biff is accelerating during this process. Cliff is on the Earth, which for the purposes of this is basically "at rest" (even though in reality the Earth moves, rotates, and accelerates in various ways). Biff is on a spaceship which undergoes intensive acceleration to read near lightspeed. This means, according to general relativity, that there are actually physical experiments that could be performed by Biff which would reveal to him that he's accelerating and the same experiments would show Cliff that he's not accelerating (or at least accelerating much less than Biff is).

The key feature is that while Cliff is in one frame of reference the entire time, Biff is actually in two frames of reference - the one where he's traveling away from the Earth and the one where he's coming back to the Earth.

So Biff's situation and Cliff's situation are *not* actually symmetrical in our scenario. Biff is absolutely the one undergoing the more significant acceleration, and therefore he's the one who undergoes the least amount of time passage.

References

- Cassidy, David C.; Holton, Gerald James; Rutherford, Floyd James (2002). Understanding Physics. Springer-Verlag. P. 422. ISBN 978-0-387-98756-9

- Mass-energy-equivalence, encyclopedia: energyeducation.ca, Retrieved 14 April, 2019

- Hands, Simon (2001). "The phase diagram of QCD". Contemporary Physics. 42 (4): 209–225. Arxiv:physics/0105022. Bibcode:2001conph..42..209H. Doi:10.1080/00107510110063843

- A-Relativity-of-Simultaneity, A-University-Physics-III-Optics-and-Modern-Physics-(openstax), A-University-Physics-(openstax), University-Physics, Bookshelves, phys.libretexts.org, Retrieved 15 May, 2019

- Landau, L.D.; Lifshitz, E.M. (2005). The Classical Theory of Fields. Course of Theoretical Physics: Volume 2. Trans. Morton Hamermesh (Fourth revised English ed.). Elsevier Butterworth-Heinemann. Pp. 116–117. ISBN 9780750627689

- Twin-paradox-real-time-travel-2699432: thoughtco.com, Retrieved 14 March, 2019

General Relativity 4

The theory of general relativity states that the observed gravitational effect between masses results from their warping of spacetime. Some of the areas which are studied in general relativity include equivalence principle, Penrose diagram, geodesics in general relativity, Mach's principle, linearized gravity, Raychaudhuri equation, etc. The diverse areas of general relativity have been thoroughly discussed in this chapter.

General Theory of Relativity

Matter does not simply pull on other matter across empty space, as Newton had imagined. Rather matter distorts space-time and it is this distorted space-time that in turn affects other matter. Objects (including planets, like the Earth, for instance) fly freely under their own inertia through warped space-time, following curved paths because this is the shortest possible path (or geodesic) in warped space-time.

This is the General Theory of Relativity, and its central premise is that the curvature of space-time is directly determined by the distribution of matter and energy contained within it. What complicates things, however, is that the distribution of matter and energy is in turn governed by the curvature of space, leading to a feedback loop and a lot of very complex mathematics. Thus, the presence of mass/energy determines the geometry of space, and the geometry of space determines the motion of mass/energy.

In practice, in our everyday world, Newton's Law of Universal Gravitation is a perfectly good approximation. The curving of light was never actually predicted by Newton but, in combination with the idea from special relativity that all forms of energy (including light) have an effective mass, then it seems logical that, as light passes a massive body like the Sun, it too will feel the tug of gravity and be bent slightly from its course. Curiously, however, Einstein's theory predicts that the path of light will be bent by twice as much as does Newton's theory, due to a kind of positive feedback. The English astronomer Arthur Eddington confirmed Einstein's predictions of the deflection of light from other stars by the Sun's gravity using measurements taken in West Africa during an eclipse of the Sun in 1919, after which the General Theory of Relativity was generally accepted in the scientific community.

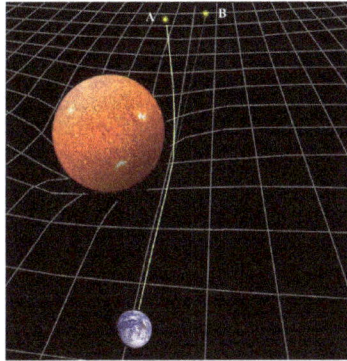

General relativity predicts the gravitational bending of light by massive bodies.

The theory has been proven remarkably accurate and robust in many different tests over the last century. The slightly elliptical orbit of planets is also explained by the theory but, even more remarkably, it also explains with great accuracy the fact that the elliptical orbits of planets are not exact repetitions but actually shift slightly with each revolution, tracing out a kind of rosette-like pattern. For instance, it correctly predicts the so-called precession of the perihelion of Mercury (that the planet Mercury traces out a complete rosette only once every 3 million years), something which Newton's Law of Universal Gravitation is not sophisticated enough to cope with.

Gravity Probe B was launched into Earth orbit in 2004, specifically to test the space-time-bending effects predicted by General Relativity using ultra-sensitive gyroscopes. The final analysis of the results in 2011 confirms the predicted effects quite closely, with a tiny 0.28% margin of error for geodetic effects and a larger 19% margin of error for the much less pronounced frame-dragging effect.

The General Theory of Relativity can actually be described using a very simple equation: R = GE (although Einstein's own formulation of his field equations are much more complex). Unfortunately, the variables in this simple equation are far from simple: R is a complicated mathematical object made up of 16 separate numbers in a matrix or "tensor" that describes the distortion of space-time; G is the gravitational constant; and E is another complicated number, also represented by a tensor, representing the energy of the object (or more accurately the 4-dimensional "energy momentum density"). Given that, though, what the equation says is simple enough: that what gravity really is is not a force but a distortion of space and time, and that the geometry of space and time depends not just on velocity but on the energy of an object. This makes sense when we consider that Newton had already shown that gravity depends on mass, and that Einstein's Special Theory of Relativity had shown that mass is equivalent to energy.

Stephen Hawking and Roger Penrose's singularity theorem of 1970 used the General Theory of Relativity to show that, just as any collapsing star must end in a singularity, the universe itself must have begun in a singularity like the Big Bang (providing that the universe does in fact contain at least as much matter as it appears to). The theorem also showed, though, that general relativity is an incomplete theory in that it cannot tell us exactly how the universe started off because it predicts that all physical theories (including itself) necessarily break down at a singularity like the Big Bang.

The theory has also provided endless fodder for the science fiction industry, predicting the existence of sci-fi staples like black holes, wormholes, time travel, parallel universes, etc. Just as an example, the notionally faster-than-light "warp" speeds of Star Trek are based firmly on relativity: if the space-time behind a starship were in some way greatly expanded, and the space-time in front of it simultaneously contracted, the starship would find itself suddenly much closer to its destination, without the local space-time around the starship being affected in any relativistic way. Unfortunately, however, such a trick would require the harvesting of vast amounts of energy, way in excess of anything imaginable today.

Equivalence Principle

In the theory of general relativity, the equivalence principle is the equivalence of gravitational and inertial mass, and Albert Einstein's observation that the gravitational "force" as experienced locally while standing on a massive body (such as the Earth) is the same as the pseudo-force experienced by an observer in a non-inertial (accelerated) frame of reference.

Einstein's Statement of the Equality of Inertial and Gravitational Mass

A little reflection will show that the law of the equality of the inertial and gravitational mass is equivalent to the assertion that the acceleration imparted to a body by a gravitational field is independent of the nature of the body. For Newton's equation of motion in a gravitational field, written out in full, it is:

(Inertial mass) . (Acceleration) . (Intensity of the gravitational field) = (Gravitational mass).

It is only when there is numerical equality between the inertial and gravitational mass that the acceleration is independent of the nature of the body.

Development of Gravitational Theory

Something like the equivalence principle emerged in the early 17th century, when Galileo expressed experimentally that the acceleration of a test mass due to gravitation is independent of the amount of mass being accelerated.

Kepler, using Galileo's discoveries, showed knowledge of the equivalence principle by accurately describing what would occur if the moon were stopped in its orbit and dropped towards Earth. This can be deduced without knowing if or in what manner gravity decreases with distance, but requires assuming the equivalency between gravity and inertia.

> "If two stones were placed in any part of the world near each other, and beyond the sphere of influence of a third cognate body, these stones, like two magnetic needles, would come together in the intermediate point, each approaching the other by a space proportional to the comparative mass of the other. If the moon and earth were not retained in their orbits by their animal force or some other equivalent, the earth would mount to the moon by a fifty-fourth part of their distance, and the moon fall towards the earth through the other fifty-three parts, and they would there meet, assuming, however, that the substance of both is of the same density."

> *— Kepler*

The 1/54 ratio is Kepler's estimate of the Moon–Earth mass ratio, based on their diameters. The accuracy of his statement can be deduced by using Newton's inertia law $F=ma$ and Galileo's gravitational observation that distance $D = (1/2)at^2$ Setting these accelerations equal for a mass is the equivalence principle. Noting the time to collision for each mass is the same gives Kepler's statement that $D_{moon}/D_{Earth} = M_{Earth}/M_{moon}$, without knowing the time to collision or how or if the acceleration force from gravity is a function of distance.

Newton's gravitational theory simplified and formalized Galileo's and Kepler's ideas by recognizing Kepler's "animal force or some other equivalent" beyond gravity and inertia were not needed, deducing from Kepler's planetary laws how gravity reduces with distance.

The equivalence principle was properly introduced by Albert Einstein in 1907, when he observed that the acceleration of bodies towards the center of the Earth at a rate of $1g$ ($g = 9.81$ m/s² being a standard reference of gravitational acceleration at the Earth's surface) is equivalent to the acceleration of an inertially moving body that would be observed on a rocket in free space being accelerated at a rate of $1g$. Einstein stated it thus:

> "We assume the complete physical equivalence of a gravitational field and a corresponding acceleration of the reference system."

> *— Einstein*

That is, being on the surface of the Earth is equivalent to being inside a spaceship (far from any sources of gravity) that is being accelerated by its engines. The direction or vector of acceleration equivalence on the surface of the earth is "up" or directly opposite the center of the planet while the vector of acceleration in a spaceship is directly opposite from the mass ejected by its thrusters. From this principle, Einstein deduced that free-fall is inertial motion. Objects in free-fall do not experience being accelerated downward (e.g. toward the earth or other massive body) but rather weightlessness and no acceleration. In an inertial frame of reference bodies (and photons, or light) obey Newton's first law, moving at constant velocity in straight lines. Analogously, in a curved spacetime the world line of an inertial particle or pulse of light is *as straight as possible* (in space and time). Such a world line is called a geodesic and from the point of view of the inertial frame is a straight line. This is why an accelerometer in free-fall doesn't register any acceleration; there isn't any.

As an example: an inertial body moving along a geodesic through space can be trapped into an orbit around a large gravitational mass without ever experiencing acceleration. This is possible because spacetime is radically curved in close vicinity to a large gravitational mass. In such a situation the geodesic lines bend inward around the center of the mass and a free-floating (weightless) inertial body will simply follow those curved geodesics into an elliptical orbit. An accelerometer on-board would never record any acceleration.

By contrast, in Newtonian mechanics, gravity is assumed to be a force. This force draws objects having mass towards the center of any massive body. At the Earth's surface, the force of gravity is counteracted by the mechanical (physical) resistance of the Earth's surface. So in Newtonian physics, a person at rest on the surface of a (non-rotating) massive object is in an inertial frame of reference. These considerations suggest the following corollary to the equivalence principle, which Einstein formulated precisely in 1911:

> "Whenever an observer detects the local presence of a force that acts on all objects in direct proportion to the inertial mass of each object, that observer is in an accelerated frame of reference."

Einstein also referred to two reference frames, K and K'. K is a uniform gravitational field, whereas K' has no gravitational field but is uniformly accelerated such that objects in the two frames experience identical forces:

> "We arrive at a very satisfactory interpretation of this law of experience, if we assume that the systems K and K' are physically exactly equivalent, that is, if we assume that we may just as well regard the system K as being in a space free from gravitational fields, if we then regard K as uniformly accelerated. This assumption of exact physical equivalence makes it impossible for us to speak of the absolute acceleration of the system of reference, just as the usual theory of relativity forbids us to talk of the absolute velocity of a system; and it makes the equal falling of all bodies in a gravitational field seem a matter of course."

— Einstein

This observation was the start of a process that culminated in general relativity. Einstein suggested that it should be elevated to the status of a general principle, which he called the "principle of equivalence" when constructing his theory of relativity:

> "As long as we restrict ourselves to purely mechanical processes in the realm where Newton's mechanics holds sway, we are certain of the equivalence of the systems K and K'. But this view of ours will not have any deeper significance unless the systems K and K' are equivalent with respect to all physical processes, that is, unless the laws of nature with respect to K are in entire agreement with those with respect to K'. By assuming this to be so, we arrive at a principle which, if it is really true, has great heuristic importance. For by theoretical consideration of processes which take place relatively to a system of reference with uniform acceleration, we obtain information as to the career of processes in a homogeneous gravitational field."

— Einstein

Einstein combined (postulated) the equivalence principle with special relativity to predict that clocks run at different rates in a gravitational potential, and light rays bend in a gravitational field, even before he developed the concept of curved spacetime.

So the original equivalence principle, as described by Einstein, concluded that free-fall and inertial motion were physically equivalent. This form of the equivalence principle can be stated as follows. An observer in a windowless room cannot distinguish between being on the surface of the Earth, and being in a spaceship in deep space accelerating at 1g. This is not strictly true, because massive bodies give rise to tidal effects (caused by variations in the strength and direction of the gravitational field) which are absent from an accelerating spaceship in deep space. The room, therefore, should be small enough that tidal effects can be neglected.

Although the equivalence principle guided the development of general relativity, it is not a founding principle of relativity but rather a simple consequence of the *geometrical* nature of the theory. In general relativity, objects in free-fall follow geodesics of spacetime, and what we perceive as the force of gravity is instead a result of our being unable to follow those geodesics of spacetime, because the mechanical resistance of matter prevents us from doing so.

Since Einstein developed general relativity, there was a need to develop a framework to test the theory against other possible theories of gravity compatible with special relativity. This was developed by Robert Dicke as part of his program to test general relativity. Two new principles were suggested, the so-called Einstein equivalence principle and the strong equivalence principle, each of which assumes the weak equivalence principle as a starting point. They only differ in whether or not they apply to gravitational experiments.

Another clarification needed is that the equivalence principle assumes a constant acceleration of 1g without considering the mechanics of generating 1g. If we do consider the mechanics of it, then we must assume the aforementioned windowless room has a fixed mass. Accelerating it at 1g means there is a constant force being applied, which = m*g where m is the mass of the windowless room along with its contents (including the observer). Now, if the observer jumps inside the room, an object lying freely on the floor will decrease in weight momentarily because the acceleration is going to decrease momentarily due to the observer pushing back against the floor in order to jump. The object will then gain weight while the observer is in the air and the resulting decreased mass of the windowless room allows greater acceleration; it will lose weight again when the observer lands and pushes once more against the floor; and it will finally return to its initial weight afterwards. To make all these effects equal those we would measure on a planet producing 1g, the windowless room must be assumed to have the same mass as that planet. Additionally, the windowless room must not cause its own gravity, otherwise the scenario changes even further. These are technicalities, clearly, but practical ones if we wish the experiment to demonstrate more or less precisely the equivalence of 1g gravity and 1g acceleration.

Modern Usage

Three forms of the equivalence principle are in current use: weak (Galilean), Einsteinian, and strong.

The Weak Equivalence Principle

The weak equivalence principle, also known as the universality of free fall or the Galilean equivalence principle can be stated in many ways. The strong EP includes (astronomic) bodies with gravitational binding energy (e.g., 1.74 solar-mass pulsar PSR J1903+0327, 15.3% of whose separated mass is absent as gravitational binding energy). The weak EP assumes falling bodies are bound by non-gravitational forces only. Either way:

- The trajectory of a point mass in a gravitational field depends only on its initial position and velocity, and is independent of its composition and *structure*.

- All test particles at the alike spacetime point, in a given gravitational field, will undergo the same acceleration, independent of their properties, including their rest mass.

- All local centers of mass free-fall (in vacuum) along identical (parallel-displaced, same speed) minimum action trajectories independent of all observable properties.

- The vacuum world-line of a body immersed in a gravitational field is independent of all observable properties.

- The local effects of motion in a curved spacetime (gravitation) are indistinguishable from those of an accelerated observer in flat spacetime, without exception.

- Mass (measured with a balance) and weight (measured with a scale) are locally in identical ratio for all bodies.

Locality eliminates measurable tidal forces originating from a radial divergent gravitational field (e.g., the Earth) upon finite sized physical bodies. The "falling" equivalence principle embraces Galileo's, Newton's, and Einstein's conceptualization. The equivalence principle does not deny the existence of measurable effects caused by a *rotating* gravitating mass (frame dragging), or bear on the measurements of light deflection and gravitational time delay made by non-local observers.

Active, Passive and Inertial Masses

By definition of active and passive gravitational mass, the force on M_1 due to the gravitational field of M_0 is:

$$F_1 = \frac{M_0^{\text{act}} M_1^{\text{pass}}}{r^2}$$

Likewise the force on a second object of arbitrary mass_2 due to the gravitational field of mass_0 is:

$$F_2 = \frac{M_0^{\text{act}} M_2^{\text{pass}}}{r^2}$$

By definition of inertial mass:

$$F = m^{\text{inert}} a$$

If m_1 and m_2 are the same distance r from m_0 then, by the weak equivalence principle, they fall at the same rate (i.e. their accelerations are the same):

$$a_1 = \frac{F_1}{m_1^{\text{inert}}} = a_2 = \frac{F_2}{m_2^{\text{inert}}}$$

Hence:

$$\frac{M_0^{act} M_1^{pass}}{r^2 m_1^{inert}} = \frac{M_0^{act} M_2^{pass}}{r^2 m_2^{inert}}$$

Therefore:

$$\frac{M_1^{pass}}{m_1^{inert}} = \frac{M_2^{pass}}{m_2^{inert}}$$

In other words, passive gravitational mass must be proportional to inertial mass for all objects.

Furthermore, by Newton's third law of motion:

$$F_1 = \frac{M_0^{act} M_1^{pass}}{r^2}$$

must be equal and opposite to:

$$F_0 = \frac{M_1^{act} M_0^{pass}}{r^2}$$

It follows that:

$$\frac{M_0^{act}}{M_0^{pass}} = \frac{M_1^{act}}{M_1^{pass}}$$

In other words, passive gravitational mass must be proportional to active gravitational mass for all objects.

The dimensionless Eötvös-parameter $\eta(A, B)$ is the difference of the ratios of gravitational and inertial masses divided by their average for the two sets of test masses "A" and "B."

$$\eta(A, B) = 2 \frac{\left(\dfrac{m_g}{m_i}\right)_A - \left(\dfrac{m_g}{m_i}\right)_B}{\left(\dfrac{m_g}{m_i}\right)_A + \left(\dfrac{m_g}{m_i}\right)_B}$$

Tests of the Weak Equivalence Principle

Tests of the weak equivalence principle are those that verify the equivalence of gravitational mass and inertial mass. An obvious test is dropping different objects, ideally in a vacuum environment, e.g., inside the Fallturm Bremen drop tower.

Experiments are still being performed at the University of Washington which have placed limits on the differential acceleration of objects towards the Earth, the Sun and towards dark matter in the galactic center. Future satellite experiments – STEP (Satellite Test of the Equivalence Principle), Galileo Galilei, and MICROSCOPE (MICROSatellite à traînée Compensée pour l'Observation du Principe d'Équivalence) – will test the weak equivalence principle in space, to much higher accuracy.

With the first successful production of antimatter, in particular anti-hydrogen, a new approach to test the weak equivalence principle has been proposed. Experiments to compare the gravitational behavior of matter and antimatter are currently being developed.

Proposals that may lead to a quantum theory of gravity such as string theory and loop quantum gravity predict violations of the weak equivalence principle because they contain many light scalar fields with long Compton wavelengths, which should generate fifth forces and variation of the fundamental constants. Heuristic arguments suggest that the magnitude of these equivalence principle violations could be in the 10^{-13} to 10^{-18} range. Currently envisioned tests of the weak equivalence principle are approaching a degree of sensitivity such that *non-discovery* of a violation would be just as profound a result as discovery of a violation. Non-discovery of equivalence principle violation in this range would suggest that gravity is so fundamentally different from other forces as to require a major reevaluation of current attempts to unify gravity with the other forces of nature. A positive detection, on the other hand, would provide a major guidepost towards unification.

The Einstein Equivalence Principle

What is now called the "Einstein equivalence principle" states that the weak equivalence principle holds, and that:

> "The outcome of any local non-gravitational experiment in a freely falling laboratory is independent of the velocity of the laboratory and its location in spacetime."

Here "local" has a very special meaning: not only must the experiment not look outside the laboratory, but it must also be small compared to variations in the gravitational field, tidal forces, so that the entire laboratory is freely falling. It also implies the absence of interactions with "external" fields other than the gravitational field.

The principle of relativity implies that the outcome of local experiments must be independent of the velocity of the apparatus, so the most important consequence of this principle is the Copernican idea that dimensionlessphysical values such as the fine-structure constant and electron-to-proton mass ratio must not depend on where in space or time we measure them. Many physicists believe that any Lorentz invariant theory that satisfies the weak equivalence principle also satisfies the Einstein equivalence principle.

Schiff's conjecture suggests that the weak equivalence principle implies the Einstein equivalence principle, but it has not been proven. Nonetheless, the two principles are tested with very different kinds of experiments. The Einstein equivalence principle has been criticized as imprecise, because there is no universally accepted way to distinguish gravitational from non-gravitational experiments.

Tests of the Einstein Equivalence Principle

In addition to the tests of the weak equivalence principle, the Einstein equivalence principle can be tested by searching for variation of dimensionless constants and mass ratios. The present best limits on the variation of the fundamental constants have mainly been set by studying the naturally occurring Oklo natural nuclear fission reactor, where nuclear reactions similar to ones we observe today have been shown to have occurred underground approximately two billion years ago. These reactions are extremely sensitive to the values of the fundamental constants.

Constant	Year	Method	Limit on fractional change
Proton gyromagnetic factor	1976	astrophysical	10^{-1}
Weak interaction constant	1976	Oklo	10^{-2}
Fine structure constant	1976	Oklo	10^{-7}
Electron−proton mass ratio	2002	quasars	10^{-4}

There have been a number of controversial attempts to constrain the variation of the strong interaction constant. There have been several suggestions that "constants" do vary on cosmological scales. The best known is the reported detection of variation (at the 10−5 level) of the fine-structure constant from measurements of distant quasars, Other researchers[who?] dispute these findings. Other tests of the Einstein equivalence principle are gravitational redshift experiments, such as the Pound–Rebka experiment which test the position independence of experiments.

The Strong Equivalence Principle

The strong equivalence principle suggests the laws of gravitation are independent of velocity and location. In particular,

> "The gravitational motion of a small test body depends only on its initial position in space-time and velocity, and not on its constitution."

and

> "The outcome of any local experiment (gravitational or not) in a freely falling laboratory is independent of the velocity of the laboratory and its location in spacetime."

The first part is a version of the weak equivalence principle that applies to objects that exert a gravitational force on themselves, such as stars, planets, black holes or Cavendish experiments. The second part is the Einstein equivalence principle (with the same definition of "local"), re-stated to allow gravitational experiments and self-gravitating bodies. The freely-falling object or laboratory, however, must still be small, so that tidal forces may be neglected (hence "local experiment").

This is the only form of the equivalence principle that applies to self-gravitating objects (such as stars), which have substantial internal gravitational interactions. It requires that the gravitational constant be the same everywhere in the universe and is incompatible with a fifth force. It is much more restrictive than the Einstein equivalence principle.

The strong equivalence principle suggests that gravity is entirely geometrical by nature (that is, the metric alone determines the effect of gravity) and does not have any extra fields associated with it. If an observer measures a patch of space to be flat, then the strong equivalence principle suggests that it is absolutely equivalent to any other patch of flat space elsewhere in the universe. Einstein's theory of general relativity (including the cosmological constant) is thought to be the only theory of gravity that satisfies the strong equivalence principle. A number of alternative theories, such as Brans–Dicke theory, satisfy only the Einstein equivalence principle.

Tests of the Strong Equivalence Principle

The strong equivalence principle can be tested by searching for a variation of Newton's gravitational constant G over the life of the universe, or equivalently, variation in the masses of the fundamental particles. A number of independent constraints, from orbits in the solar system and studies of Big Bang nucleosynthesis have shown that G cannot have varied by more than 10%.

Thus, the strong equivalence principle can be tested by searching for fifth forces (deviations from the gravitational force-law predicted by general relativity). These experiments typically look for failures of the inverse-square law (specifically Yukawa forces or failures of Birkhoff's theorem) behavior of gravity in the laboratory. The most accurate tests over short distances have been performed by the Eöt–Wash group. A future satellite experiment, SEE (Satellite Energy Exchange), will search for fifth forces in space and should be able to further constrain violations of the strong equivalence principle. Other limits, looking for much longer-range forces, have been placed by searching for the Nordtvedt effect, a "polarization" of solar system orbits that would be caused by gravitational self-energy accelerating at a different rate from normal matter. This effect has been sensitively tested by the Lunar Laser Ranging Experiment. Other tests include studying the deflection of radiation from distant radio sources by the sun, which can be accurately measured by very long baseline interferometry. Another sensitive test comes from measurements of the frequency shift of signals to and from the Cassini spacecraft. Together, these measurements have put tight limits on Brans–Dicke theory and other alternative theories of gravity.

In 2014, astronomers discovered a stellar triple system including a millisecond pulsar PSR J0337+1715 and two white dwarfs orbiting it. The system provided them a chance to test the strong equivalence principle in a strong gravitational field with high accuracy.

Challenges

One challenge to the equivalence principle is the Brans–Dicke theory. Self-creation cosmology is a modification of the Brans–Dicke theory. The Fredkin Finite Nature Hypothesis is an even more radical challenge to the equivalence principle and has even fewer supporters.

In August 2010, researchers from the University of New South Wales, Swinburne University of Technology, and Cambridge University published a paper titled "Evidence for spatial variation of the fine structure constant", whose tentative conclusion is that, "qualitatively, the results suggest a violation of the Einstein Equivalence Principle, and could infer a very large or infinite universe, within which our 'local' Hubble volume represents a tiny fraction."

Dutch physicist and string theorist Erik Verlinde has generated a self-contained, logical derivation of the equivalence principle based on the starting assumption of a holographic universe. Given this situation, gravity would not be a true fundamental force as is currently thought but instead an "emergent property" related to entropy. Verlinde's entropic gravity theory apparently leads naturally to the correct observed strength of dark energy; previous failures to explain its incredibly small magnitude have been called by such people as cosmologist Michael Turner (who is credited as having coined the term "dark energy") as "the greatest embarrassment in the history of theoretical physics". These ideas are far from settled and still very controversial.

Penrose Diagram

We can't directly visualize a four-dimensional manifold. When a spacetime has a symmetry, however, we may be able to visualize the relevant properties the whole thing by considering a lower-dimensional part of it. By analogy, if we wanted to visualize the structure of the earth's interior, we might draw a diagram showing a two-dimensional section through its center. In fact, we could get rid of two dimensions and simply draw a diagram of a single radial line running from the earth's core to its surface; each point on this line would then represent a sphere. If we do this in general relativity, for a spacetime that is spherically symmetric, then we can reduce the four-dimensional to a two-dimensional one, with each point representing a two-sphere. By applying some further tricks, we will see that we can end up with a very convenient and useful visualization called a Penrose diagram, also known as a Penrose-Carter diagram or causal diagram.

Flat Spacetime

As a warmup, figure shows a Penrose diagram for flat (Minkowski) spacetime. The diagram looks $1 + 1$-dimensional, but the convention is that spherical symmetry is assumed, so two more dimensions are hidden, and we're really portraying $3 + 1$ dimensions. A typical point on the interior of the diamond region represents a 2-sphere. On this type of diagram, light cones look just like they would on a normal spacetime diagram of Minkowski space, but distance scales are highly distorted. The diamond represents the entire spacetime, with the distortion fitting this entire infinite region into that finite area. Despite the distortion, the diagram shows lightlike surfaces as 45-degree diagonals. Spacelike and timelike geodesics, however, are distorted, as shown by the curves in the diagram.

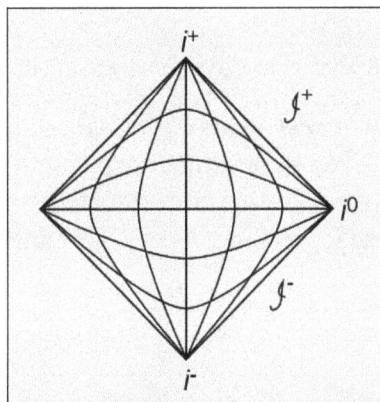

Penrose diagram for flat spacetime.

The distortion becomes greater as we move away from the center of the diagram, and becomes infinite near the edges. Because of this infinite distortion, the points i^- and i^+ actually represent 3-spheres. All timelike curves start at i^- and end at i^+, which are idealized points at infinity, like the vanishing points in perspective drawings. We can think of i^+ as the "Elephants' graveyard," where massive particles go when they die. Similarly, lightlike curves end on $\mathcal{I}^+$, referred to as *null infinity*. The point at i^0 is an infinitely distant endpoint for spacelike curves. Because of the spherical symmetry, the left and right halves of the diagram are redundant.

It is possible to make up explicit formulae that translate back and forth between Minkowski coordinates and points on the diamond, but in general this is not necessary. In fact, the utility of the diagrams is that they let us think about causal relationships in coordinate-independent ways. A light cone on the diagram looks exactly like a normal light cone.

Since this particular spacetime is homogeneous, it makes no difference what spatial location on the diagram we pick as our axis of symmetry. For example, we could arbitrarily pick the left-hand corner, the central timelike geodesic (drawn straight) or one of the other timelike geodesics (represented as if it were curved).

Schwarzschild Spacetime

Figure is a Penrose diagram for the Schwarzschild spacetime, i.e., a spacetime that looks like Minkowski space, except that it has one eternal black hole in it. This is a black hole that did not form by gravitational collapse. This spacetime isn't homogeneous; it has a specific location that is its center of spherical symmetry, and this is the vertical line on the left marked r = 0. The triangle is the spacetime inside the event horizon; we could have copied it across the r = 0 line if we had so desired, but the copies would have been redundant.

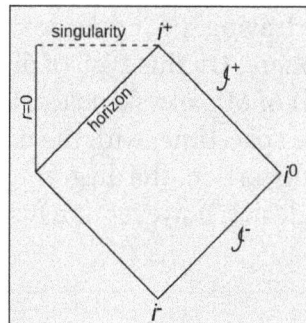

Penrose diagram for Schwarzschild spacetime: a black hole that didn't form by gravitational collapse.

The Penrose diagram makes it easy to reason about causal relationships. For example, we can see that if a particle reaches a point inside the event horizon, its entire causal future lies inside the horizon, and all of its possible future world-lines intersect the singularity. The horizon is a lightlike surface, which makes sense, because it's defined as the boundary of the set of points from which a light ray could reach J^+.

Astrophysical Black Hole

Figure is a Penrose diagram for a black hole that has formed by gravitational collapse. Using this type of diagram, we can succinctly address one of the most vexing FAQs about black holes. If a distant observer watches the collapsing cloud of matter from which the black hole forms, her optical observations will show that the light from the matter becomes more and more gravitationally redshifted, and if she wishes, she can interpret this as an example of gravitational time dilation. As she waits longer and longer, the light signals from the infalling matter take longer and longer to arrive. The redshift approaches infinity as the matter approaches the horizon, so the light waves ultimately become too low in energy to be detectable by any given instrument. Furthermore, her patience (or her lifetime) will run out, because the time on her clock approaches infinity as she waits to get signals from matter that is approaching the horizon. This is all exactly as it should be, since the horizon is by definition the boundary of her observable universe. (A light ray emitted from the horizon will end up at i+, which is an end-point of timelike world-lines reached only by observers who have experienced an infinite amount of proper time).

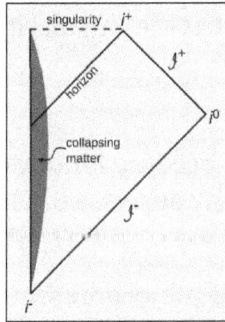

Penrose diagram for a black hole formed by gravitational collapse.

People who are bothered by these issues often acknowledge the external *unobservability* of matter passing through the horizon, and then want to pass from this to questions like, "Does that mean the black hole never really forms?" This presupposes that our distant observer has a uniquely defined notion of simultaneity that applies to a region of space stretching from her own position to the interior of the black hole, so that she can say what's going on inside the black hole "now." But the notion of simultaneity in general relativity is even more limited than its counterpart in special relativity. Not only is simultaneity in general relativity observer-dependent, as in special relativity, but it is also local rather than global.

In figure, E is an event on the world-line of an observer. The spacelike surface S_1 is one possible "now" for this observer. According to this surface, no particle has ever fallen in and reached the horizon; every such particle has a world-line that intersects S_1, and therefore it's still on its way in.

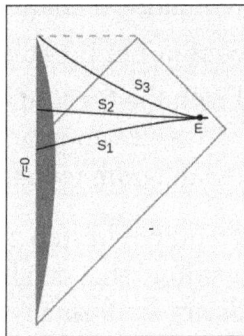

S_2 is another possible "now" for the same observer at the same time. According to this definition of "now," all the particles have passed the event horizon, but none have hit the singularity yet. Finally, S_3 is a "now" according to which all the particles have hit the singularity.

If this was special relativity, then we could decide which surface was the correct notion of simultaneity for the observer, based on the observer's state of motion. But in general relativity, this only works locally. There is no well-defined way of deciding which is the correct way of globally extending this notion of simultaneity.

Although it may seem strange that we can't say whether the singularity has "already" formed according to a distant observer, this is really just an inevitable result of the fact that the singularity is spacelike. The same thing happens in the case of a Schwarzschild spacetime, which we think of as a description of an eternal black hole, i.e., one that has always existed and always will. On the similar Penrose diagram for an eternal black hole, we can still draw a spacelike surface like S_1 or S_2, representing a definition of "now" such that the singularity doesn't exist yet.

Penrose Diagrams in General

Ideally we would like to generalize the procedure for drawing Penrose diagrams so that we would be able to uniquely determine one for any spacetime. This turns out to be not so clear-cut. The procedure would go something like this:

1. Make an n-dimensional section or projection, where usually, but not always, n = 2.

2. Do a transformation to reduce the resulting manifold to a flat one of finite size.

3. Adjoin idealized surfaces and points at infinity.

At step 1, we want to take advantage of any symmetries, such as rotational symmetry, so that the final result will be informative, be representative of the whole spacetime, and accurately depict causal relationships in the original spacetime. If the original spacetime has a low degree of symmetry (e.g., a spacetime containing three black holes arranged in a triangle), then this might require n > 2. At this step we also need to make sure that lightlike geodesics in the original space correspond properly to lightlike geodesics in the submanifold.

For step 2, we have already given a geometrical characterization of the type of transformation we have in mind, which is called a *conformal transformation*. It turns out to be possible to encapsulate this idea in a simple analytic way. Given a spacetime with a metric g, we define a fictitious metric $\tilde{g} = \Omega^2 g$, , where Ω is a nonzero real number that varies from point to point. The idea here is that g and $\tilde{g}$ agree on where the light cone is, but they disagree on the measurement of distances and times. The same manifold equipped with the fictitious metric $\tilde{g}$ is the one being drawn when we make a Penrose diagram. We let $\Omega \to 0$ as we approach the idealized boundary regions like i^o and J^+, and this is what causes the Penrose diagram to take up finite space.

It is not possible in general to do what is required in step 2 by making a conformal transformation to change a manifold into a flat one. A manifold that can be flattened in this way is called conformally flat. All two-dimensional manifolds are conformally flat, so in the n = 2 case this is guaranteed. For n > 3 we will usually not have conformal flatness if there are gravitational waves or tidal forces present.

Global Hyperbolicity

Causality refers to our vaguely defined feeling that the world should have an orderly progression of cause and effect. Making this notion more precise is surprisingly difficult. Penrose diagrams, and their associated concepts, are essentially representations of the causal structure of spacetime, and these turn out to be helpful in putting together one of the more satisfying attempts to define causality. This definition is called global hyperbolicity.

Geodesics in General Relativity

In general relativity, a geodesic generalizes the notion of a "straight line" to curved spacetime. Importantly, the world line of a particle free from all external, non-gravitational force is a particular type of geodesic. In other words, a freely moving or falling particle always moves along a geodesic.

In general relativity, gravity can be regarded as not a force but a consequence of a curved space-time geometry where the source of curvature is the stress–energy tensor (representing matter, for instance). Thus, for example, the path of a planet orbiting a star is the projection of a geodesic of the curved 4-D spacetime geometry around the star onto 3-D space.

Mathematical Expression

The full geodesic equation is:

$$\frac{d^2 x^\mu}{ds^2} + \Gamma^\mu{}_{\alpha\beta} \frac{dx^\alpha}{ds} \frac{dx^\beta}{ds} = 0$$

where s is a scalar parameter of motion (e.g. the proper time), and $\Gamma^\mu{}_{\alpha\beta}$ are Christoffel symbols (sometimes called the affine connection coefficients or Levi-Civita connection coefficients) which is symmetric in the two lower indices. Indices may take the values: 0, 1, 2, 3 and the summation convention is used for repeated indices α and β. The quantity on the left-hand-side of this equation is the acceleration of a particle, and so this equation is analogous to Newton's laws of motion which likewise provide formulae for the acceleration of a particle. This equation of motion employs the Einstein notation, meaning that repeated indices are summed (i.e. from zero to three). The Christoffel symbols are functions of the four space-time coordinates, and so are independent of the velocity or acceleration or other characteristics of a test particlewhose motion is described by the geodesic equation.

Equivalent Mathematical Expression using Coordinate Time as Parameter

So far the geodesic equation of motion has been written in terms of a scalar parameter s. It can alternatively be written in terms of the time coordinate, $t \equiv x^0$ (here we have used the triple bar to signify a definition). The geodesic equation of motion then becomes:

$$\frac{d^2 x^\mu}{dt^2} = -\Gamma^\mu{}_{\alpha\beta} \frac{dx^\alpha}{dt} \frac{dx^\beta}{dt} + \Gamma^0{}_{\alpha\beta} \frac{dx^\alpha}{dt} \frac{dx^\beta}{dt} \frac{dx^\mu}{dt}.$$

This formulation of the geodesic equation of motion can be useful for computer calculations and to compare General Relativity with Newtonian Gravity. It is straightforward to derive this form of the geodesic equation of motion from the form which uses proper time as a parameter, using the chain rule. Notice that both sides of this last equation vanish when the mu index is set to zero. If the particle's velocity is small enough, then the geodesic equation reduces to this:

$$\frac{d^2 x^n}{dt^2} = -\Gamma^n{}_{00}.$$

Here the index n takes the values (1,2,3). This equation simply means that all test particles at a particular place and time will have the same acceleration, which is a well-known feature of Newtonian gravity. For example, everything floating around in the international space station will undergo roughly the same acceleration due to gravity.

Derivation Directly from the Equivalence Principle

Physicist Steven Weinberg has presented a derivation of the geodesic equation of motion directly from the equivalence principle. The first step in such a derivation is to suppose that no particles are accelerating in the neighborhood of a point-event with respect to a freely falling coordinate system (X^μ). Setting $T \equiv X^0$, we have the following equation that is locally applicable in free fall:

$$\frac{d^2 X^\mu}{dT^2} = 0.$$

The next step is to employ the multi-dimensional chain rule. We have:

$$\frac{dX^\mu}{dT} = \frac{dx^\nu}{dT} \frac{\partial X^\mu}{\partial x^\nu}$$

Differentiating once more with respect to the time, we have:

$$\frac{d^2 X^\mu}{dT^2} = \frac{d^2 x^\nu}{dT^2} \frac{\partial X^\mu}{\partial x^\nu} + \frac{dx^\nu}{dT} \frac{dx^\alpha}{dT} \frac{\partial^2 X^\mu}{\partial x^\nu \partial x^\alpha}$$

Therefore:

$$\frac{d^2 x^\nu}{dT^2} \frac{\partial X^\mu}{\partial x^\nu} = -\frac{dx^\nu}{dT} \frac{dx^\alpha}{dT} \frac{\partial^2 X^\mu}{\partial x^\nu \partial x^\alpha}$$

Multiply both sides of this last equation by the following quantity:

$$\frac{\partial x^\lambda}{\partial X^\mu}$$

Consequently, we have this:

$$\frac{d^2 x^\lambda}{dT^2} = -\frac{dx^\nu}{dT} \frac{dx^\alpha}{dT} \left[\frac{\partial^2 X^\mu}{\partial x^\nu \partial x^\alpha} \frac{\partial x^\lambda}{\partial X^\mu} \right].$$

Using:

$$\Gamma^{\lambda}{}_{v\alpha} = \left[\frac{\partial^2 X^{\mu}}{\partial x^{v} \partial x^{\alpha}} \frac{\partial x^{\lambda}}{\partial X^{\mu}} \right]$$

it becomes:

$$\frac{d^2 x^{\lambda}}{dT^2} = -\Gamma^{\lambda}{}_{v\alpha} \frac{dx^{v}}{dT} \frac{dx^{\alpha}}{dT}.$$

Applying the one-dimensional chain rule gives:

$$\frac{d^2 x^{\lambda}}{dt^2} \left(\frac{dt}{dT} \right)^2 + \frac{dx^{\lambda}}{dt} \frac{d^2 t}{dT^2} = -\Gamma^{\lambda}{}_{v\alpha} \frac{dx^{v}}{dt} \frac{dx^{\alpha}}{dt} \left(\frac{dt}{dT} \right)^2.$$

$$\frac{d^2 x^{\lambda}}{dt^2} + \frac{dx^{\lambda}}{dt} \frac{d^2 t}{dT^2} \left(\frac{dT}{dt} \right)^2 = -\Gamma^{\lambda}{}_{v\alpha} \frac{dx^{v}}{dt} \frac{dx^{\alpha}}{dt}.$$

As before, we can set $t \equiv x^0$. Then the first derivative of x^0 with respect to t is one and the second derivative is zero. Replacing λ with zero gives:

$$\frac{d^2 t}{dT^2} \left(\frac{dT}{dt} \right)^2 = -\Gamma^{0}{}_{v\alpha} \frac{dx^{v}}{dt} \frac{dx^{\alpha}}{dt}.$$

Subtracting $d x^{\lambda} / d t$ times this from the previous equation gives:

$$\frac{d^2 x^{\lambda}}{dt^2} = -\Gamma^{\lambda}{}_{v\alpha} \frac{dx^{v}}{dt} \frac{dx^{\alpha}}{dt} + \Gamma^{0}{}_{v\alpha} \frac{dx^{v}}{dt} \frac{dx^{\alpha}}{dt} \frac{dx^{\lambda}}{dt}$$

which is a form of the geodesic equation of motion (using the coordinate time as parameter).

The geodesic equation of motion can alternatively be derived using the concept of parallel transport.

Deriving the Geodesic Equation via an Action

We can (and this is the most common technique) derive the geodesic equation via the action principle. Consider the case of trying to find a geodesic between two timelike-separated events.

Let the action be:

$$S = \int ds$$

where $ds = \sqrt{-g_{\mu v}(x) dx^{\mu} dx^{v}}$ is the line element. There is a negative sign inside the square root because the curve must be timelike. To get the geodesic equation we must vary this action. To do this let us parameterize this action with respect to a parameter λ. Doing this we get:

$$S = \int \sqrt{-g_{\mu v} \frac{dx^{\mu}}{d\lambda} \frac{dx^{v}}{d\lambda}} \, d\lambda$$

We can now go ahead and vary this action with respect to the curve x^μ. By the principle of least action we get:

$$0 = \delta S = \int \delta \left(\sqrt{-g_{\mu\nu} \frac{dx^\mu}{d\lambda} \frac{dx^\nu}{d\lambda}} \right) d\lambda = \int \frac{\delta \left(-g_{\mu\nu} \frac{dx^\mu}{d\lambda} \frac{dx^\nu}{d\lambda} \right)}{2\sqrt{-g_{\mu\nu} \frac{dx^\mu}{d\lambda} \frac{dx^\nu}{d\lambda}}} d\lambda$$

Using the product rule we get:

$$0 = \int \left(\frac{dx^\mu}{d\lambda} \frac{dx^\nu}{d\tau} \delta g_{\mu\nu} + g_{\mu\nu} \frac{d\delta x^\mu}{d\lambda} \frac{dx^\nu}{d\tau} + g_{\mu\nu} \frac{dx^\mu}{d\tau} \frac{d\delta x^\nu}{d\lambda} \right) d\lambda = \int \left(\frac{dx^\mu}{d\lambda} \frac{dx^\nu}{d\tau} \partial_\alpha g_{\mu\nu} \delta x^\alpha + 2 g_{\mu\nu} \frac{d\delta x^\mu}{d\lambda} \frac{dx^\nu}{d\tau} \right) d\lambda$$

Integrating by-parts the last term and dropping the total derivative (which equals to zero at the boundaries) we get that:

$$0 = \int \left(\frac{dx^\mu}{d\tau} \frac{dx^\nu}{d\tau} \partial_\alpha g_{\mu\nu} \delta x^\alpha - 2\delta x^\mu \frac{d}{d\tau} \left(g_{\mu\nu} \frac{dx^\nu}{d\tau} \right) \right) d\tau = \int \left(\frac{dx^\mu}{d\tau} \frac{dx^\nu}{d\tau} \partial_\alpha g_{\mu\nu} \delta x^\alpha - 2\delta x^\mu \partial_\alpha g_{\mu\nu} \frac{dx^\alpha}{d\tau} \frac{dx^\nu}{d\tau} - 2\delta x^\mu g_{\mu\nu} \frac{d^2 x^\nu}{d\tau^2} \right) d\tau$$

Simplifying a bit we see that:

$$0 = \int \left(-2 g_{\mu\nu} \frac{d^2 x^\nu}{d\tau^2} + \frac{dx^\alpha}{d\tau} \frac{dx^\nu}{d\tau} \partial_\mu g_{\alpha\nu} - 2 \frac{dx^\alpha}{d\tau} \frac{dx^\nu}{d\tau} \partial_\alpha g_{\mu\nu} \right) \delta x^\mu d\tau$$

so,

$$0 = \int \left(-2 g_{\mu\nu} \frac{d^2 x^\nu}{d\tau^2} + \frac{dx^\alpha}{d\tau} \frac{dx^\nu}{d\tau} \partial_\mu g_{\alpha\nu} - \frac{dx^\alpha}{d\tau} \frac{dx^\nu}{d\tau} \partial_\alpha g_{\mu\nu} - \frac{dx^\nu}{d\tau} \frac{dx^\alpha}{d\tau} \partial_\nu g_{\mu\alpha} \right) \delta x^\mu d\tau$$

multiplying this equation by $-\dfrac{1}{2}$ we get:

$$0 = \int \left(g_{\mu\nu} \frac{d^2 x^\nu}{d\tau^2} + \frac{1}{2} \frac{dx^\alpha}{d\tau} \frac{dx^\nu}{d\tau} \left(\partial_\alpha g_{\mu\nu} + \partial_\nu g_{\mu\alpha} - \partial_\mu g_{\alpha\nu} \right) \right) \delta x^\mu d\tau$$

So by Hamilton's principle we find that the Euler–Lagrange equation is:

$$g_{\mu\nu} \frac{d^2 x^\nu}{d\tau^2} + \frac{1}{2} \frac{dx^\alpha}{d\tau} \frac{dx^\nu}{d\tau} \left(\partial_\alpha g_{\mu\nu} + \partial_\nu g_{\mu\alpha} - \partial_\mu g_{\alpha\nu} \right) = 0$$

Multiplying by the inverse metric tensor $g^{\mu\beta}$ we get that:

$$\frac{d^2 x^\beta}{d\tau^2} + \frac{1}{2} g^{\mu\beta} \left(\partial_\alpha g_{\mu\nu} + \partial_\nu g_{\mu\alpha} - \partial_\mu g_{\alpha\nu} \right) \frac{dx^\alpha}{d\tau} \frac{dx^\nu}{d\tau} = 0$$

Thus we get the geodesic equation:

$$\frac{d^2 x^\beta}{d\tau^2} + \Gamma^\beta_{\ \alpha v} \frac{dx^\alpha}{d\tau} \frac{dx^v}{d\tau} = 0$$

with the Christoffel symbol defined in terms of the metric tensor as:

$$\Gamma^\beta_{\ \alpha v} = \frac{1}{2} g^{\mu\beta} \left(\partial_\alpha g_{\mu v} + \partial_v g_{\mu\alpha} - \partial_\mu g_{\alpha v} \right)$$

Similar derivations, with minor amendments, can be used to produce analogous results for geodesics between light-like or space-like separated pairs of points.

Equation of Motion

Albert Einstein believed that the geodesic equation of motion can be derived from the field equations for empty space, i.e. from the fact that the Ricci curvature vanishes. He wrote:

> "It has been shown that this law of motion — generalized to the case of arbitrarily large gravitating masses — can be derived from the field equations of empty space alone. According to this derivation the law of motion is implied by the condition that the field be singular nowhere outside its generating mass points."

and

> "One of the imperfections of the original relativistic theory of gravitation was that as a field theory it was not complete; it introduced the independent postulate that the law of motion of a particle is given by the equation of the geodesic."

A complete field theory knows only fields and not the concepts of particle and motion. For these must not exist independently from the field but are to be treated as part of it.

On the basis of the description of a particle without singularity, one has the possibility of a logically more satisfactory treatment of the combined problem: The problem of the field and that of the motion coincide.

Both physicists and philosophers have often repeated the assertion that the geodesic equation can be obtained from the field equations to describe the motion of a gravitational singularity, but this claim remains disputed. Less controversial is the notion that the field equations determine the motion of a fluid or dust, as distinguished from the motion of a point-singularity.

Extension to the Case of a Charged Particle

In deriving the geodesic equation from the equivalence principle, it was assumed that particles in a local inertial coordinate system are not accelerating. However, in real life, the particles may be charged, and therefore may be accelerating locally in accordance with the Lorentz force. That is:

$$\frac{d^2 X^\mu}{ds^2} = \frac{q}{m} F^{\mu\beta} \frac{dX^\alpha}{ds} \eta_{\alpha\beta}.$$

With:

$$\eta_{\alpha\beta}\frac{dX^{\alpha}}{ds}\frac{dX^{\beta}}{ds}=-1.$$

The Minkowski tensor $\eta_{\alpha\beta}$ is given by:

$$\eta_{\alpha\beta}=\begin{pmatrix}-1 & 0 & 0 & 0\\ 0 & 1 & 0 & 0\\ 0 & 0 & 1 & 0\\ 0 & 0 & 0 & 1\end{pmatrix}$$

These last three equations can be used as the starting point for the derivation of an equation of motion in General Relativity, instead of assuming that acceleration is zero in free fall. Because the Minkowski tensor is involved here, it becomes necessary to introduce something called the metric tensor in General Relativity. The metric tensor g is symmetric, and locally reduces to the Minkowski tensor in free fall. The resulting equation of motion is as follows:

$$\frac{d^{2}x^{\mu}}{ds^{2}}=-\Gamma^{\mu}{}_{\alpha\beta}\frac{dx^{\alpha}}{ds}\frac{dx^{\beta}}{ds}+\frac{q}{m}F^{\mu\beta}\frac{dx^{\alpha}}{ds}g_{\alpha\beta}.$$

with:

$$g_{\alpha\beta}\frac{dx^{\alpha}}{ds}\frac{dx^{\beta}}{ds}=-1.$$

This last equation signifies that the particle is moving along a timelike geodesic; massless particles like the photon instead follow null geodesics (replace −1 with zero on the right-hand side of the last equation). It is important that the last two equations are consistent with each other, when the latter is differentiated with respect to proper time, and the following formula for the Christoffel symbols ensures that consistency:

$$\Gamma^{\lambda}{}_{\alpha\beta}=\frac{1}{2}g^{\lambda\tau}\left(\frac{\partial g_{\tau\alpha}}{\partial x^{\beta}}+\frac{\partial g_{\tau\beta}}{\partial x^{\alpha}}-\frac{\partial g_{\alpha\beta}}{\partial x^{\tau}}\right)$$

This last equation does not involve the electromagnetic fields, and it is applicable even in the limit as the electromagnetic fields vanish. The letter g with superscripts refers to the inverse of the metric tensor. In General Relativity, indices of tensors are lowered and raised by contraction with the metric tensor or its inverse, respectively.

Geodesics as Curves of Stationary Interval

A geodesic between two events can also be described as the curve joining those two events which has a stationary interval (4-dimensional "length"). *Stationary* here is used in the sense in which that term is used in the calculus of variations, namely, that the interval along the curve varies minimally among curves that are nearby to the geodesic.

In Minkowski space there is only one time-like geodesic that connects any given pair of time-like separated events, and that geodesic is the curve with the longest proper time between the two events. But in curved spacetime, it's possible for a pair of widely separated events to have more than one time-like geodesic that connects them. In such instances, the proper times along the various geodesics will not in general be the same. And for some geodesics in such instances, it's possible for a curve that connects the two events and is nearby to the geodesic to have either a longer or a shorter proper time than the geodesic.

For a space-like geodesic through two events, there are always nearby curves which go through the two events that have either a longer or a shorter proper length than the geodesic, even in Minkowski space. In Minkowski space, in an inertial frame of reference in which the two events are simultaneous, the geodesic will be the straight line between the two events at the time at which the events occur. Any curve that differs from the geodesic purely spatially (*i.e.* does not change the time coordinate) in that frame of reference will have a longer proper length than the geodesic, but a curve that differs from the geodesic purely temporally (*i.e.* does not change the space coordinate) in that frame of reference will have a shorter proper length.

The interval of a curve in spacetime is:

$$l = \int \sqrt{\left| g_{\mu\nu} \dot{x}^\mu \dot{x}^\nu \right|} \, ds \, .$$

Then, the Euler–Lagrange equation,

$$\frac{d}{ds} \frac{\partial}{\partial \dot{x}^\alpha} \sqrt{\left| g_{\mu\nu} \dot{x}^\mu \dot{x}^\nu \right|} = \frac{\partial}{\partial x^\alpha} \sqrt{\left| g_{\mu\nu} \dot{x}^\mu \dot{x}^\nu \right|} \, ,$$

becomes, after some calculation,

$$2(\Gamma^\lambda_{\ \mu\nu} \dot{x}^\mu \dot{x}^\nu + \ddot{x}^\lambda) = U^\lambda \frac{d}{ds} \ln |U_\nu U^\nu| \, ,$$

where $U^\mu = \dot{x}^\mu$.

Proof:

If the parameter s is chosen to be affine, then the right side of the above equation vanishes (because $U_\nu U^\nu$ is constant). Finally, we have the geodesic equation:

$$\Gamma^\lambda_{\ \mu\nu} \dot{x}^\mu \dot{x}^\nu + \ddot{x}^\lambda = 0 \, .$$

Mach's Principle

In theoretical physics, particularly in discussions of gravitation theories, Mach's principle (or Mach's conjecture) is the name given by Einstein to an imprecise hypothesis often credited to the physicist and philosopher Ernst Mach. The idea is that the existence of absolute rotation (the distinction of local inertial frames vs. rotating reference frames) is determined by the large-scale distribution of matter, as exemplified by this anecdote:

> "You are standing in a field looking at the stars. Your arms are resting freely at your side, and you see that the distant stars are not moving. Now start spinning. The stars are whirling around you and your arms are pulled away from your body. Why should your arms be pulled away when the stars are whirling? Why should they be dangling freely when the stars don't move?"

Mach's principle says that this is not a coincidence—that there is a physical law that relates the motion of the distant stars to the local inertial frame. If you see all the stars whirling around you, Mach suggests that there is some physical law which would make it so you would feel a centrifugal force. There are a number of rival formulations of the principle. It is often stated in vague ways, like "mass out there influences inertia here". A very general statement of Mach's principle is "local physical laws are determined by the large-scale structure of the universe".

This concept was a guiding factor in Einstein's development of the general theory of relativity. Einstein realized that the overall distribution of matter would determine the metric tensor, which tells you which frame is rotationally stationary. Frame-dragging and conservation of gravitational angular momentum makes this into a true statement in the general theory in certain solutions. But because the principle is so vague, many distinct statements can be (and have been) made that would qualify as a Mach principle, and some of these are false. The Gödel rotating universe is a solution of the field equations that is designed to disobey Mach's principle in the worst possible way. In this example, the distant stars seem to be revolving faster and faster as one moves further away. This example doesn't completely settle the question, because it has closed timelike curves.

Einstein's use of the Principle

There is a fundamental issue in relativity theory. If all motion is relative, how can we measure the inertia of a body? We must measure the inertia with respect to something else. But what if we imagine a particle completely on its own in the universe? We might hope to still have some notion of its state of motion. Mach's principle is sometimes interpreted as the statement that such a particle's state of motion has no meaning in that case.

In Mach's words, the principle is embodied as follows:

> "[The] investigator must feel the need of knowledge of the immediate connections, say, of the masses of the universe. There will hover before him as an ideal insight into the principles of the whole matter, from which accelerated and inertial motions will result in the same way. "

Albert Einstein seemed to view Mach's principle as something along the lines of:

> "inertia originates in a kind of interaction between bodies"

In this sense, at least some of Mach's principles are related to philosophical holism. Mach's suggestion can be taken as the injunction that gravitation theories should be relational theories. Einstein brought the principle into mainstream physics while working on general relativity. Indeed, it was Einstein who first coined the phrase *Mach's principle*. There is much debate as to whether Mach really intended to suggest a new physical law since he never states it explicitly.

The writing in which Einstein found inspiration from Mach was "The Science of Mechanics", where the philosopher criticized Newton's idea of absolute space, in particular the argument that Newton gave sustaining the existence of an advantaged reference system: what is commonly called "Newton's bucket argument".

In his *Philosophiae Naturalis Principia Mathematica*, Newton tried to demonstrate that:

> "one can always decide if one is rotating with respect to the absolute space, measuring the apparent forces that arise only when an absolute rotation is performed. If a bucket is filled with water, and made to rotate, initially the water remains still, but then, gradually, the walls of the vessel communicate their motion to the water, making it curve and climb up the borders of the bucket, because of the centrifugal forces produced by the rotation. This thought experiment demonstrates that the centrifugal forces arise only when the water is in rotation with respect to the absolute space (represented here by the earth's reference frame, or better, the distant stars) instead, when the bucket was rotating with respect to the water no centrifugal forces were produced, this indicating that the latter was still with respect to the absolute space."

Mach, in his book, says that:

> "the bucket experiment only demonstrates that when the water is in rotation with respect to the bucket no centrifugal forces are produced, and that we cannot know how the water would behave if in the experiment the bucket's walls were increased in depth and width until they became leagues big. In Mach's idea this concept of absolute motion should be substituted with a total relativism in which every motion, uniform or accelerated, has sense only in reference to other bodies (*i.e.*, one cannot simply say that the water is rotating, but must specify if it's rotating with respect to the vessel or to the earth). In this view, the apparent forces that seem to permit discrimination between relative and "absolute" motions should only be considered as an effect of the particular asymmetry that there is in our reference system between the bodies which we consider in motion, that are small (like buckets), and the bodies that we believe are still (the earth and distant stars), that are overwhelmingly bigger and heavier than the former."

This same thought had been expressed by the philosopher George Berkeley in his *De Motu*. It is then not clear, in the passages from Mach just mentioned, if the philosopher intended to formulate a new kind of physical action between heavy bodies. This physical mechanism should determine the inertia of bodies, in a way that the heavy and distant bodies of our universe should contribute the most to the inertial forces. More likely, Mach only suggested a mere "redescription of motion in space as experiences that do not invoke the term *space*". What is certain is that Einstein interpreted Mach's passage in the former way, originating a long-lasting debate.

Most physicists believe Mach's principle was never developed into a quantitative physical theory that would explain a mechanism by which the stars can have such an effect. It was never made clear by Mach himself exactly what his principle was. Although Einstein was intrigued and inspired by Mach's principle, Einstein's formulation of the principle is not a fundamental assumption of general relativity.

Mach's Principle in General Relativity

Because intuitive notions of distance and time no longer apply, what exactly is meant by "Mach's principle" in general relativity is even less clear than in Newtonian physics and at least 21 formulations of Mach's principle are possible, some being considered more strongly Machian than others. A relatively weak formulation is the assertion that the motion of matter in one place should affect which frames are inertial in another.

Einstein, before completing his development of the general theory of relativity, found an effect which he interpreted as being evidence of Mach's principle. We assume a fixed background for conceptual simplicity, construct a large spherical shell of mass, and set it spinning in that background. The reference frame in the interior of this shell will precess with respect to the fixed background. This effect is known as the Lense–Thirring effect. Einstein was so satisfied with this manifestation of Mach's principle that he wrote a letter to Mach expressing this:

> "it turns out that inertia originates in a kind of interaction between bodies, quite in the sense of your considerations on Newton's pail experiment If one rotates [a heavy shell of matter] relative to the fixed stars about an axis going through its center, a Coriolis force arises in the interior of the shell; that is, the plane of a Foucault pendulum is dragged around (with a practically unmeasurably small angular velocity)."

The Lense–Thirring effect certainly satisfies the very basic and broad notion that "matter there influences inertia here". The plane of the pendulum would not be dragged around if the shell of matter were not present, or if it were not spinning. As for the statement that "inertia originates in a kind of interaction between bodies", this too could be interpreted as true in the context of the effect.

More fundamental to the problem, however, is the very existence of a fixed background, which Einstein describes as "the fixed stars". Modern relativists see the imprints of Mach's principle in the initial-value problem. Essentially, we humans seem to wish to separate spacetime into slices of constant time. When we do this, Einstein's equations can be decomposed into one set of equations, which must be satisfied on each slice, and another set, which describe how to move between slices. The equations for an individual slice are elliptic partial differential equations. In general, this means that only part of the geometry of the slice can be given by the scientist, while the geometry everywhere else will then be dictated by Einstein's equations on the slice.

In the context of an asymptotically flat spacetime, the boundary conditions are given at infinity. Heuristically, the boundary conditions for an asymptotically flat universe define a frame with respect to which inertia has meaning. By performing a Lorentz transformation on the distant universe, of course, this inertia can also be transformed.

A stronger form of Mach's principle applies in Wheeler–Mach–Einstein spacetimes, which require spacetime to be spatially compact and globally hyperbolic. In such universes Mach's principle can be stated as *the distribution of matter and field energy-momentum (and possibly other information) at a particular moment in the universe determines the inertial frame at each point in the universe* (where "a particular moment in the universe" refers to a chosen Cauchy surface).

There have been other attempts to formulate a theory that is more fully Machian, such as the Brans–Dicke theory and the Hoyle–Narlikar theory of gravity, but most physicists argue that none have been fully successful. At an exit poll of experts, held in Tübingen in 1993, when asked the question "Is general relativity perfectly Machian?", 3 respondents replied "yes", and 22 replied "no". To the question "Is general relativity with appropriate boundary conditions of closure of some kind very Machian?" the result was 14 "yes" and 7 "no".

However, Einstein was convinced that a valid theory of gravity would necessarily have to include the relativity of inertia:

> "So strongly did Einstein believe at that time in the relativity of inertia that in 1918 he stated as being on an equal footing three principles on which a satisfactory theory of gravitation should rest:
>
> • The principle of relativity as expressed by general covariance.
>
> • The principle of equivalence.
>
> • Mach's principle (the first time this term entered the literature): that the $g_{\mu\nu}$ are completely determined by the mass of bodies, more generally by $T_{\mu\nu}$.
>
> In 1922, Einstein noted that others were satisfied to proceed without this [third] criterion and added, "This contentedness will appear incomprehensible to a later generation however."
>
> It must be said that, as far as I can see, to this day, Mach's principle has not brought physics decisively farther. It must also be said that the origin of inertia is and remains the most obscure subject in the theory of particles and fields. Mach's principle may therefore have a future – but not without the quantum theory."
>
> *—Abraham Pais*

Variations in the Statement of the Principle

The broad notion that "mass there influences inertia here" has been expressed in several forms. Hermann Bondi and Joseph Samuel have listed eleven distinct statements that can be called Mach principles, labelled by *Mach0* through *Mach10*. Though their list is not necessarily exhaustive, it does give a flavor for the variety possible.

• *Mach0*: The universe, as represented by the average motion of distant galaxies, does not appear to rotate relative to local inertial frames.

• *Mach1*: Newton's gravitational constant G is a dynamical field.

• *Mach2*: An isolated body in otherwise empty space has no inertia.

• *Mach3*: Local inertial frames are affected by the cosmic motion and distribution of matter.

• *Mach4*: The universe is spatially closed.

- *Mach5*: The total energy, angular and linear momentum of the universe are zero.

- *Mach6*: Inertial mass is affected by the global distribution of matter.

- *Mach7*: If you take away all matter, there is no more space.

- *Mach8*: $\Omega \overset{\text{def}}{=} 4\pi\rho GT^2$ is a definite number, of order unity, where ρ is the mean density of matter in the universe, and T is the Hubble time.

- *Mach9*: The theory contains no absolute elements.

- *Mach10*: Overall rigid rotations and translations of a system are unobservable.

ADM Formalism

The ADM formalism (named for its authors Richard Arnowitt, Stanley Deser and Charles W. Misner) is a Hamiltonian formulation of general relativity that plays an important role in canonical quantum gravity and numerical relativity. It was first published in 1959.

The formalism supposes that spacetime is foliated into a family of spacelike surfaces Σ_t, labeled by their time coordinate t, and with coordinates on each slice given by x^i. The dynamic variables of this theory are taken to be the metric tensor of three dimensional spatial slices $\gamma_{ij}(t, x^k)$ and their conjugate momenta $\pi^{ij}(t, x^k)$. Using these variables it is possible to define a Hamiltonian, and thereby write the equations of motion for general relativity in the form of Hamilton's equations.

In addition to the twelve variables γ_{ij} and π^{ij}, there are four Lagrange multipliers: the lapse function, N, and components of shift vector field, N_i. These describe how each of the "leaves" Σ_t of the foliation of spacetime are welded together. The equations of motion for these variables can be freely specified; this freedom corresponds to the freedom to specify how to lay out the coordinate system in space and time.

Most references adopt notation in which four dimensional tensors are written in abstract index notation, and that Greek indices are spacetime indices taking values (0, 1, 2, 3) and Latin indices are spatial indices taking values (1, 2, 3). In the derivation here, a superscript (4) is prepended to quantities that typically have both a three-dimensional and a four-dimensional version, such as the metric tensor for three-dimensional slices g_{ij} and the metric tensor for the full four-dimensional spacetime $^{(4)}g_{\mu\nu}$.

The text here uses Einstein notation in which summation over repeated indices is assumed.

Two types of derivatives are used: Partial derivatives are denoted either by the operator ∂_i or by subscripts preceded by a comma. Covariant derivatives are denoted either by the operator ∇_i or by subscripts preceded by a semicolon.

The absolute value of the determinant of the matrix of metric tensor coefficients is represented by g (with no indices). Other tensor symbols written without indices represent the trace of the corresponding tensor such as $\pi = g^{ij}\pi_{ij}$.

Derivation

Lagrangian Formulation

The starting point for the ADM formulation is the Lagrangian:

$$\mathcal{L} = {}^{(4)}R\sqrt{(4)g},$$

which is a product of the square root of the determinant of the four-dimensional metric tensor for the full spacetime and its Ricci scalar. This is the Lagrangian from the Einstein–Hilbert action.

The desired outcome of the derivation is to define an embedding of three-dimensional spatial slices in the four-dimensional spacetime. The metric of the three-dimensional slices:

$$g_{ij} = {}^{(4)}g_{ij}$$

will be the generalized coordinates for a Hamiltonian formulation. The conjugate momenta can then be computed as:

$$\pi^{ij} = \sqrt{{}^{(4)}g}\left({}^{(4)}\Gamma^0_{pq} - g_{pq}{}^{(4)}\Gamma^0_{rs}g^{rs} \right)g^{ip}g^{jq},$$

using standard techniques and definitions. The symbols ${}^{(4)}\Gamma^0_{ij}$ are Christoffel symbols associated with the metric of the full four-dimensional spacetime. The lapse:

$$N = \left(-{}^{(4)}g^{00} \right)^{-1/2}$$

and the shift vector:

$$N_i = {}^{(4)}g_{0i}$$

are the remaining elements of the four-metric tensor.

Having identified the quantities for the formulation, the next step is to rewrite the Lagrangian in terms of these variables. The new expression for the Lagrangian:

$$\mathcal{L} = -g_{ij}\partial_t\pi^{ij} - NH - N_iP^i - 2\partial_i\left(\pi^{ij}N_j - \frac{1}{2}\pi N^i + \nabla^i N\sqrt{g} \right)$$

is conveniently written in terms of the two new quantities:

$$H = -\sqrt{g}\left[{}^{(3)}R + g^{-1}\left(\frac{1}{2}\pi^2 - \pi^{ij}\pi_{ij} \right) \right]$$

and

$$P^i = -2\pi^{ij}{}_{;j},$$

which are known as the Hamiltonian constraint and the momentum constraint respectively. Note also that the lapse and the shift appear in the Hamiltonian as Lagrange multipliers.

Equations of Motion

Although the variables in the Lagrangian represent the metric tensor on three-dimensional spaces embedded in the four-dimensional spacetime, it is possible and desirable to use the usual procedures from Lagrangian mechanics to derive "equations of motion" that describe the time evolution of both the metric g_{ij} and its conjugate momentum π^{ij}.

$$\partial_t g_{ij} = \frac{2N}{\sqrt{g}}\left(\pi_{ij} - \tfrac{1}{2}\pi g_{ij}\right) + N_{i;j} + N_{j;i}$$

and

$$\partial_t \pi^{ij} = -N\sqrt{g}\left(R^{ij} - \tfrac{1}{2}Rg^{ij}\right) + \frac{N}{2\sqrt{g}}g^{ij}\left(\pi^{mn}\pi_{mn} - \tfrac{1}{2}\pi^2\right) - \frac{2N}{\sqrt{g}}\left(\pi^{in}\pi_n{}^j - \tfrac{1}{2}\pi\pi^{ij}\right)$$
$$- \sqrt{g}\left(\nabla^i\nabla^j N - g^{ij}\nabla^n\nabla_n N\right) + \nabla_n\left(\pi^{ij}N^n\right) - N^i{}_{;n}\pi^{nj} - N^j{}_{;n}\pi^{ni}$$

is a non-linear set of partial differential equations.

Taking variations with respect to the lapse and shift provide constraint equations:

$$H = 0$$

and

$$P^i = 0,$$

and the lapse and shift themselves can be freely specified, reflecting the fact that coordinate systems can be freely specified in both space and time.

Applications

Application to Quantum Gravity

Using the ADM formulation, it is possible to attempt to construct a quantum theory of gravity in the same way that one constructs the Schrödinger equation corresponding to a given Hamiltonian in quantum mechanics. That is, replace the canonical momenta $\pi^{ij}(t, x^k)$ and the spatial metric functions by linear functional differential operators:

$$\hat{g}_{ij}(t, x^k) \mapsto g_{ij}(t, x^k),$$

$$\hat{\pi}^{ij}(t, x^k) \mapsto -i\frac{\delta}{\delta g_{ij}(t, x^k)}.$$

More precisely, the replacing of classical variables by operators is restricted by commutation relations. The hats represents operators in quantum theory. This leads to the Wheeler–DeWitt equation.

Application to Numerical Solutions of the Einstein Equations

There are relatively few known exact solutions to the Einstein field equations. In order to find other solutions, there is an active field of study known as numerical relativity in which supercomputers are used to find approximate solutions to the equations. In order to construct such solutions numerically, most researchers start with a formulation of the Einstein equations closely related to the ADM formulation. The most common approaches start with an initial value problem based on the ADM formalism.

In Hamiltonian formulations, the basic point is replacement of set of second order equations by another first order set of equations. We may get this second set of equations by Hamiltonian formulation in an easy way. Of course this is very useful for numerical physics, because the reduction of order of differential equations must be done, if we want to prepare equations for a computer.

ADM Energy and Mass

ADM energy is a special way to define the energy in general relativity, which is only applicable to some special geometries of spacetime that asymptotically approach a well-defined metric tensor at infinity – for example a spacetime that asymptotically approaches Minkowski space. The ADM energy in these cases is defined as a function of the deviation of the metric tensor from its prescribed asymptotic form. In other words, the ADM energy is computed as the strength of the gravitational field at infinity.

If the required asymptotic form is time-independent (such as the Minkowski space itself), then it respects the time-translational symmetry. Noether's theorem then implies that the ADM energy is conserved. According to general relativity, the conservation law for the total energy does not hold in more general, time-dependent backgrounds – for example, it is completely violated in physical cosmology. Cosmic inflation in particular is able to produce energy (and mass) from "nothing" because the vacuum energy density is roughly constant, but the volume of the Universe grows exponentially.

Application to Modified Gravity

By using the ADM decomposition and introducing extra auxiliary fields, in 2009 Deruelle et al. found a method to find the Gibbons–Hawking–York boundary term for modified gravity theories "whose Lagrangian is an arbitrary function of the Riemann tensor".

Linearized Gravity

In the theory of general relativity, linearized gravity is the application of perturbation theory to the metric tensor that describes the geometry of spacetime. As a consequence, linearized gravity is an effective method for modeling the effects of gravity when the gravitational field is weak. The usage of linearized gravity is integral to the study of gravitational waves and weak-field gravitational lensing.

Weak-field Approximation

The Einstein field equation (EFE) describing the geometry of spacetime is given as (using Natural units):

$$R_{\mu\nu} - \frac{1}{2}Rg_{\mu\nu} = 8\pi G T_{\mu\nu}$$

where $R_{\mu\nu}$ is the Ricci tensor, R is the Ricci scalar, $T_{\mu\nu}$ is the energy-momentum tensor, and $g_{\mu\nu}$ is the spacetime metric tensor that represent the solutions of the equation.

Although succinct when written out using Einstein notation, hidden within the Ricci tensor and Ricci scalar are exceptionally nonlinear dependencies on the metric which render the prospect of finding exact solutions impractical in most systems. However, when describing particular systems for which the curvature of spacetime is small (meaning that terms in the EFE that are quadratic in $g_{\mu\nu}$ do not significantly contribute to the equations of motion), one can model the solution of the field equations as being the Minkowski metric $\eta_{\mu\nu}$ plus a small perturbation term $h_{\mu\nu}$. In other words:

$$g_{\mu\nu} = \eta_{\mu\nu} + h_{\mu\nu}, \qquad |h_{\mu\nu}| \ll 1.$$

In this regime, substituting the general metric $g_{\mu\nu}$ for this perturbative approximation results in a simplified expression for the Ricci tensor:

$$R_{\mu\nu} = \frac{1}{2}(\partial_\sigma\partial_\mu h_\nu^\sigma + \partial_\sigma\partial_\nu h_\mu^\sigma - \partial_\mu\partial_\nu h - \Box h_{\mu\nu}),$$

where $h = \eta^{\mu\nu}h_{\mu\nu}$ is the trace of the perturbation, ∂_μ denotes the partial derivative with respect to the x^μ coordinate of spacetime, and $\Box = \eta^{\mu\nu}\partial_\mu\partial_\nu$ is the d' Alembert operator.

Together with the Ricci scalar,

$$R = \eta_{\mu\nu}R^{\mu\nu} = \partial_\mu\partial_\nu h^{\mu\nu} - \Box h,$$

the left side of the field equation reduces to:

$$R_{\mu\nu} - \frac{1}{2}Rg_{\mu\nu} = \frac{1}{2}(\partial_\sigma\partial_\mu h_\nu^\sigma + \partial_\sigma\partial_\nu h_\mu^\sigma - \partial_\mu\partial_\nu h - \Box h_{\mu\nu} - \eta_{\mu\nu}\partial_\rho\partial_\lambda h^{\rho\lambda} + \eta_{\mu\nu}\Box h).$$

and thus the EFE is reduced to a linear, second order partial differential equation in terms of $h_{\mu\nu}$.

Gauge Invariance

The process of decomposing the general spacetime $g_{\mu\nu}$ into the Minkowski metric plus a perturbation term is not unique. This is due to the fact that different choices for coordinates may give different forms for $_{\mu\nu}$. In order to capture this phenomena, the application of gauge symmetry is applied.

Gauge symmetries are a mathematical device for describing a system that does not change when the underlying coordinate system is "shifted" by an infinitesimal amount. So although the perturbation metric $h_{\mu\nu}$ is not consistently defined between different coordinate systems, the overall system which it describes *is*.

To capture this formally, the non-uniqueness of the perturbation $h_{\mu\nu}$ is represented as being a consequence of the diverse collection of diffeomorphisms on spacetime that leave $h_{\mu\nu}$ sufficiently small. Therefore to continue, it is required that $h_{\mu\nu}$ be defined in terms of a general set of diffeomorphisms then select the subset of these that preserve the small scale that is required by the weak-field approximation. One may thus define ϕ to denote an arbitrary diffeomorphism that maps the flat Minkowski spacetime to the more general spacetime represented by the metric $g_{\mu\nu}$. With this, the perturbation metric may be defined as the difference between the pullback of $g_{\mu\nu}$ and the Minkowski metric:

$$h_{\mu\nu} = (\phi^* g)_{\mu\nu} - \eta_{\mu\nu}.$$

The diffeomorphisms ϕ may thus be chosen such that $|h_{\mu\nu}| \ll 1$.

Given then a vector field ξ^μ defined on the flat, background spacetime, an additional family of diffeomorphisms ψ_ϵ may be defined as those generated by ξ^μ and parameterized by $\epsilon > 0$. These new diffeomorphisms will be used to represent the coordinate transformations for "infinitesimal shifts" as discussed above. Together with ϕ, a family of perturbations is given by:

$$h^{(\epsilon)}_{\mu\nu} = [(\phi \circ \psi_\epsilon)^* g]_{\mu\nu} - \eta_{\mu\nu}$$
$$= [\psi_\epsilon^* (\phi^* g)]_{\mu\nu} - \eta_{\mu\nu}$$
$$= \psi_\epsilon^* (h + \eta)_{\mu\nu} - \eta_{\mu\nu}$$
$$= (\psi_\epsilon^* h)_{\mu\nu} + \epsilon \left[\frac{(\psi_\epsilon^* \eta)_{\mu\nu} - \eta_{\mu\nu}}{\epsilon} \right].$$

Therefore, in the limit $\epsilon \to 0$,

$$h^{(\epsilon)}_{\mu\nu} = h_{\mu\nu} + \epsilon \mathcal{L}_\xi \eta_{\mu\nu}$$

where $\mathcal{L}_\xi$ is the Lie derivative along the vector field ξ_μ.

The Lie derivative works out to yield the final gauge transformation of the perturbation metric $h_{\mu\nu}$:

$$h^{(\epsilon)}_{\mu\nu} = h_{\mu\nu} + \epsilon(\partial_\mu \xi_\nu + \partial_\nu \xi_\mu),$$

which precisely define the set of perturbation metrics that describe the same physical system. In other words, it characterizes the gauge symmetry of the linearized field equations.

Choice of Gauge

By exploiting gauge invariance, certain properties of the perturbation metric can be guaranteed by choosing a suitable vector field ξ^μ.

Transverse Gauge

To study how the perturbation $h_{\mu\nu}$ distorts measurements of length, it is useful to define the following spatial tensor:

$$s_{ij} = h_{ij} - \frac{1}{3}\delta^{kl}h_{kl}\delta_{ij}$$

The indices span only spatial components: $i, j \in \{1, 2, 3\}$) Thus, by using s_{ij} the spatial components of the perturbation can be decomposed as:

$$h_{ij} = s_{ij} - \Psi\delta_{ij}$$

where $\Psi = -\frac{1}{3}\delta^{kl}h_{kl}$.

The tensor s_{ij} is, by construction, traceless and is referred to as the strain since it represents the amount by which the perturbation stretches and contracts measurements of space. In the context of studying gravitational radiation, the strain is particularly useful when utilized with the transverse gauge. This gauge is defined by choosing the spatial components of ξ^{μ} to satisfy the relation:

$$\nabla^2\xi^j + \frac{1}{3}\partial_j\partial_i\xi^i = -\partial_i s^{ij},$$

then choosing the time component ξ^0 to satisfy:

$$\nabla^2\xi^0 = \partial_i h_{0i} + \partial_0\partial_i\xi^i.$$

After performing the gauge transformation using the formula, the strain becomes spatially transverse:

$$\partial_i s^{ij}_{(\epsilon)} = 0,$$

with the additional property:

$$\partial_i h^{0i}_{(\epsilon)} = 0.$$

Synchronous Gauge

The synchronous gauge simplifies the perturbation metric by requiring that the metric not distort measurements of time. More precisely, the synchronous gauge is chosen such that the non-spatial components of $h^{(\epsilon)}_{\mu\nu}$ are zero, namely:

$$h^{(\epsilon)}_{0\nu} = 0.$$

This can be achieved by requiring the time component of ξ^{μ} to satisfy:

$$\partial_0\xi^0 = -h_{00}$$

and requiring the spatial components to satisfy:

$$\partial_0 \xi^i = \partial_i \xi^0 - h_{0i}.$$

Harmonic Gauge

The *harmonic gauge* (also referred to as the *Lorenz gauge*) is selected whenever it is necessary to reduce the linearized field equations as much as possible. This can be done if the condition:

$$\partial_\mu h_\nu^\mu = \frac{1}{2} \partial_\nu h$$

is true. To achieve this, ξ_μ is required to satisfy the relation:

$$\Box \xi_\mu = -\partial_\nu h_\mu^\nu + \frac{1}{2} \partial_\mu h.$$

Consequently, by using the harmonic gauge, the Einstein tensor $G_{\mu\nu} = R_{\mu\nu} - \frac{1}{2} R g_{\mu\nu}$

$$G_{\mu\nu} = -\frac{1}{2} \Box \left(h_{\mu\nu}^{(\epsilon)} - \frac{1}{2} h^{(\epsilon)} \eta_{\mu\nu} \right).$$

Therefore, by writing it in terms of a "trace-reversed" metric, $\bar{h}_{\mu\nu}^{(\epsilon)} = h_{\mu\nu}^{(\epsilon)} - \frac{1}{2} h^{(\epsilon)} \eta_{\mu\nu}$ the linearized field equations reduce to:

$$\Box \bar{h}_{\mu\nu}^{(\epsilon)} = -16\pi G T_{\mu\nu}.$$

Which can be solved exactly using the wave solutions that define gravitational radiation.

Raychaudhuri Equation

In general relativity, the Raychaudhuri equation, or Landau–Raychaudhuri equation, is a fundamental result describing the motion of nearby bits of matter.

The equation is important as a fundamental lemma for the Penrose-Hawking singularity theorems and for the study of exact solutions in general relativity, but has independent interest, since it offers a simple and general validation of our intuitive expectation that gravitation should be a universal attractive force between any two bits of mass-energy in general relativity, as it is in Newton's theory of gravitation.

The equation was discovered independently by the Indian physicist Amal Kumar Raychaudhuri and the Soviet physicist Lev Landau.

Given a timelike unit vector field $\vec{X}$ (which can be interpreted as a family or congruence of nonintersecting world lines via the integral curve, not necessarily geodesics), Raychaudhuri's equation can be written:

$$\dot{\theta} = -\frac{\theta^2}{3} - 2\sigma^2 + 2\omega^2 - E[\vec{X}]^a_{\ a} + \dot{X}^a_{;a}$$

where:

$$2\sigma^2 = \sigma_{mn}\,\sigma^{mn},\, 2\omega^2 = \omega_{mn}\,\omega^{mn}$$

are (non-negative) quadratic invariants of the shear tensor:

$$\sigma_{ab} = \theta_{ab} - \frac{1}{3}\theta\,h_{ab}$$

and the vorticity tensor:

$$\omega_{ab} = h^m_{\ a}\,h^n_{\ b}X_{[m;n]}$$

respectively. Here,

$$\theta_{ab} = h^m_{\ a}\,h^n_{\ b}X_{(m;n)}$$

is the expansion tensor, θ is its trace, called the expansion scalar, and:

$$h_{ab} = g_{ab} + X_a\,X_b$$

is the projection tensor onto the hyperplanes orthogonal to $\vec{X}$ Also, dot denotes differentiation with respect to proper time counted along the world lines in the congruence. Finally, the trace of the tidal tensor $E[\vec{X}]_{ab}$ can also be written:

$$E[\vec{X}]^a_{\ a} = R_{mn}\,X^m\,X^n + 1$$

This quantity is sometimes called the Raychaudhuri scalar.

Intuitive Significance

The expansion scalar measures the fractional rate at which the volume of a small ball of matter changes with respect to time as measured by a central comoving observer (and so it may take negative values). In other words, the above equation gives us the evolution equation for the expansion of the timelike congruence. If the derivative (with respect to proper time) of this quantity turns out to be *negative* along some world line (after a certain event), then any expansion of a small ball of matter (whose center of mass follows the world line in question) must be followed by recollapse. If not, continued expansion is possible.

The shear tensor measures any tendency of an initially spherical ball of matter to become distorted into an ellipsoidal shape. The vorticity tensor measures any tendency of nearby world lines to twist about one another (if this happens, our small blob of matter is rotating, as happens to fluid elements in an ordinary fluid flow which exhibits nonzero vorticity).

The right hand side of Raychaudhuri's equation consists of two types of terms:

- Terms which promote (re)-collapse.

  ◦ Initially nonzero expansion scalar.

  ◦ Nonzero shearing.

  ◦ Positive trace of the tidal tensor; this is precisely the condition guaranteed by assuming the *strong energy condition*, which holds for the most important types of solutions, such as physically reasonable fluid solutions).

- Terms which oppose (re)-collapse:

  ◦ Nonzero vorticity, corresponding to Newtonian centrifugal forces.

  ◦ Positive divergence of the acceleration vector (e.g., outward pointing acceleration due to a spherically symmetric explosion, or more prosaically, due to body forces on fluid elements in a ball of fluid held together by its own self-gravitation).

Usually one term will win out. However, there are situations in which a balance can be achieved. This balance may be:

- Stable: In the case of hydrostatic equilibrium of a ball of perfect fluid (e.g. in a model of a stellar interior), the expansion, shear, and vorticity all vanish, and a radial divergence in the acceleration vector (the necessary body force on each blob of fluid being provided by the pressure of surrounding fluid) counteracts the Raychaudhuri scalar, which for a perfect fluid is $\pi\mu$. In Newtonian gravitation, the trace of the tidal tensor is $E[\vec{X}]_{ab} = 4\pi(\mu+3p)$; in general relativity, the tendency of pressure to oppose gravity is partially offset by this term, which under certain circumstances can become important.

- Unstable: For example, the world lines of the dust particles in the Gödel solution have vanishing shear, expansion, and acceleration, but constant vorticity just balancing a constant Raychuadhuri scalar due to nonzero vacuum energy ("cosmological constant").

Focusing Theorem

Suppose the strong energy condition holds in some region of our spacetime, and let $\vec{X}$ be a timelike *geodesic* unit vector field with *vanishing vorticity*, or equivalently, which is hypersurface orthogonal. For example, this situation can arise in studying the world lines of the dust particles in cosmological models which are exact dust solutions of the Einstein field equation (provided that these world lines are not twisting about one another, in which case the congruence would have nonzero vorticity).

Then Raychaudhuri's equation becomes:

$$\dot{\theta} = -\frac{\theta^2}{3} - 2\sigma^2 - E[\vec{X}]^a_{\ a}$$

Now the right hand side is always negative, so even if the expansion scalar is initially positive (if our small ball of dust is initially increasing in volume), eventually it must become negative (our ball of dust must recollapse).

Indeed, in this situation we have:

$$\dot{\theta} \leq -\frac{\theta^2}{3}$$

Integrating this inequality with respect to proper time τ gives:

$$\frac{1}{\theta} \geq \frac{1}{\theta_0} + \frac{\tau}{3}$$

If the initial value θ_0 of the expansion scalar is negative, this means that our geodesics must converge in a caustic (θ goes to minus infinity) within a proper time of at most $-3/\theta_0$ after the measurement of the initial value θ_0 of the expansion scalar. This need not signal an encounter with a curvature singularity, but it does signal a breakdown in our mathematical description of the motion of the dust.

Optical Equations

There is also an optical (or null) version of Raychaudhuri's equation for null geodesic congruences.

$$\dot{\hat{\theta}} = -\frac{1}{2}\hat{\theta}^2 - 2\hat{\sigma}^2 + 2\hat{\omega}^2 - T_{\mu\nu}U^\mu U^\nu.$$

Here, the hats indicate that the expansion, shear and vorticity are only with respect to the transverse directions. When the vorticity is zero, then assuming the null energy condition, caustics will form before the affine parameter reaches $2/\hat{\theta}_0$.

Applications

The event horizon is defined as the boundary of the causal past of null infinity. Such boundaries are generated by null geodesics. The affine parameter goes to infinity as we approach null infinity, and no caustics form until then. So, the expansion of the event horizon has to be nonnegative. As the expansion gives the rate of change of the logarithm of the area density, this means the event horizon area can never go down, at least classically, assuming the null energy condition.

Hamilton–Jacobi–Einstein Equation

In general relativity, the Hamilton–Jacobi–Einstein equation (HJEE) or Einstein–Hamilton–Jacobi equation (EHJE) is an equation in the Hamiltonian formulation of geometrodynamics in superspace, cast in the "geometrodynamics era" around the 1960s, by Asher Peres in 1962 and others. It is an attempt to reformulate general relativity in such a way that it resembles quantum theory within a semiclassical approximation, much like the correspondence between quantum mechanics and classical mechanics.

It is named for Albert Einstein, Carl Gustav Jacob Jacobi, and William Rowan Hamilton. The EHJE contains as much information as all ten Einstein field equations (EFEs). It is a modification of the Hamilton–Jacobi equation (HJE) from classical mechanics, and can be derived from the Einstein–Hilbert action using the principle of least action in the ADM formalism.

Correspondence between Classical and Quantum Physics

In classical analytical mechanics, the dynamics of the system is summarized by the action S. In quantum theory, namely non-relativistic quantum mechanics (QM), relativistic quantum mechanics (RQM), as well as quantum field theory (QFT), with varying interpretations and mathematical formalisms in these theories, the behavior of a system is completely contained in a complex-valued probability amplitude Ψ (more formally as a quantum state ket $|\Psi\rangle$ - an element of a Hilbert space). Using the polar form of the wave function, so making a Madelung transformation:

$$\Psi = \sqrt{\rho}e^{iS/\hbar}$$

the phase of Ψ is interpreted as the action, and the modulus $\sqrt{\rho} = \sqrt{\Psi^*\Psi} = |\Psi|$ is interpreted according to the Copenhagen interpretation as the probability density function. The reduced Planck constant $\hbar$ is the *quantum of angular momentum*. Substitution of this into the quantum general Schrödinger equation (SE):

$$i\hbar\frac{\partial\Psi}{\partial t} = \hat{H}\Psi,$$

and taking the limit $\hbar \to 0$ yields the classical HJE:

$$-\frac{\partial S}{\partial t} = H,$$

which is one aspect of the correspondence principle.

Shortcomings of Four-dimensional Spacetime

On the other hand, the transition between quantum theory and general relativity (GR) is difficult to make; one reason is the treatment of space and time in these theories. In non-relativistic QM, space and time are not on equal footing; time is a parameter while position is an operator. In RQM and QFT, position returns to the usual spatial coordinates alongside the time coordinate, although these theories are consistent only with SR in four-dimensional flat Minkowski space, and not curved space nor GR. It is possible to formulate quantum field theory in curved spacetime, yet even this still cannot incorporate GR because gravity is not renormalizable in QFT. Additionally, in GR particles move through curved spacetime with a deterministically known position and momentum at every instant, while in quantum theory, the position and momentum of a particle cannot be exactly known simultaneously; space x and momentum p, and energy E and time t, are pairwise subject to the uncertainty principles.

$$\Delta x \Delta p \geq \frac{\hbar}{2}, \quad \Delta E \Delta t \geq \frac{\hbar}{2},$$

which imply that small intervals in space and time mean large fluctuations in energy and momentum are possible. Since in GR mass–energy and momentum–energy is the source of spacetime curvature, large fluctuations in energy and momentum mean the spacetime "fabric" could potentially become so distorted that it breaks up at sufficiently small scales. There is theoretical and experimental evidence from QFT that vacuum does have energy since the motion of electrons in atoms is fluctuated, this is related to the Lamb shift. For these reasons and others, at increasingly small scales, space and time are thought to be dynamical up to the Planck length and Planck time scales.

In any case, a four-dimensional curved spacetime continuum is a well-defined and central feature of general relativity, but not in quantum mechanics.

One attempt to find an equation governing the dynamics of a system, in as close a way as possible to QM and GR, is to reformulate the HJE in *three-dimensional curved space* understood to be "dynamic" (changing with time), and *not* four-dimensional spacetime dynamic in all four dimensions, as the EFEs are. The space has a metric.

The metric tensor in general relativity is an essential object, since proper time, arc length, geodesic motion in curved spacetime, and other things, all depend on the metric. The HJE above is modified to include the metric, although it's only a function of the 3d spatial coordinates r, (for example r = (x, y, z) in Cartesian coordinates) without the coordinate time *t*:

$$g_{ij} = g_{ij}(\mathbf{r}).$$

In this context g_{ij} is referred to as the "metric field" or simply "field".

General Equation (Free Curved Space)

For a free particle in curved "empty space" or "free space", i.e. in the absence of matter other than the particle itself, the equation can be written:

$$\frac{1}{\sqrt{g}}\left(\frac{1}{2}g_{pq}g_{rs} - g_{pr}g_{qs}\right)\frac{\delta S}{\delta g_{pq}}\frac{\delta S}{\delta g_{rs}} + \sqrt{g}R = 0$$

where g is the determinant of the metric tensor and R the Ricci scalar curvature of the 3d geometry (not including time), and the "δ" instead of "d" denotes the variational derivative rather than the ordinary derivative. These derivatives correspond to the field momenta "conjugate to the metric field":

$$\pi^{ij}(\mathbf{r}) = \pi^{ij} = \frac{\delta S}{\delta g_{ij}},$$

the rate of change of action with respect to the field coordinates $g_{ij}(\mathbf{r})$. The g and π here are analogous to q and $p = \partial S/\partial q$, respectively, in classical Hamiltonian mechanics.

The equation describes how wavefronts of constant action propagate in superspace - as the dynamics of matter waves of a free particle unfolds in curved space. Additional source terms are needed to account for the presence of extra influences on the particle, which include the presence of other particles or distributions of matter (which contribute to space curvature), and sources of electromagnetic fields affecting particles with electric charge or spin. Like the Einstein field equations, it is non-linear in the metric because of the products of the metric components, and like the HJE it is non-linear in the action due to the product of variational derivatives in the action.

The quantum mechanical concept, that action is the phase of the wavefunction, can be interpreted from this equation as follows. The phase has to satisfy the principle of least action; it must be stationary for a small change in the configuration of the system, in other words for a slight change in the position of the particle, which corresponds to a slight change in the metric components;

$$g_{ij} \rightarrow g_{ij} + \delta g_{ij},$$

the slight change in phase is zero:

$$\delta S = \int \frac{\delta S}{\delta g_{ij}(\mathbf{r})} \delta g_{ij}(\mathbf{r}) \mathrm{d}^3 \mathbf{r} = 0,$$

(where d³r is the volume element of the volume integral). So the constructive interference of the matter waves is a maximum. This can be expressed by the superposition principle; applied to many non-localized wavefunctions spread throughout the curved space to form a localized wavefunction:

$$\Psi = \sum_n c_n \psi_n,$$

for some coefficients c_n, and additionally the action (phase) S_n for each ψ_n must satisfy:

$$\delta S = S_{n+1} - S_n = 0,$$

for all n, or equivalently,

$$S_1 = S_2 = \cdots = S_n = \cdots$$

Regions where Ψ is maximal or minimal occur at points where there is a probability of finding the particle there, and where the action (phase) change is zero. So in the EHJE above, each wavefront of constant action is where the particle could be found.

This equation still does not "unify" quantum mechanics and general relativity, because the semi-classical Eikonal approximation in the context of quantum theory and general relativity has been applied, to provide a transition between these theories.

Applications

The equation takes various complicated forms in:

- Quantum gravity
- Quantum cosmology

Einstein Field Equations

The Einstein Field Equation (EFE) is also known as Einstein's equation. There are a set of ten equations extracted from Albert Einstein's General Theory of Relativity. The EFE describes the basic interaction of gravitation. The equations were first published in 1915 by Albert Einstein as a tensor equation.

Following is the Einstein Field Equation:

$$G_{\mu\nu} + g_{\mu\nu}\Lambda = \frac{8\pi G}{c^4}T_{\mu\nu}$$

where,

- $G_{\mu\nu}$ is the Einstein tensor which is given as $R\mu\nu-\frac{1}{2} Rg\mu\nu$
- $R_{\mu\nu}$ is the Ricci curvature tensor
- R is the scalar curvature
- $g_{\mu\nu}$ is the metric tensor
- Λ is a cosmological constant
- G is Newton's gravitational constant
- c is the speed of light
- $T_{\mu\nu}$ is the stress-energy tensor

Einstein Field Equations Derivation

Following is the derivation of Einstein Field Equations. Einstein wanted to explain that measure of curvature = source of gravity.

The source of gravity is the stress-energy tensor. The stress-energy tensor is given as:

$$T^{\alpha\beta} = \begin{bmatrix} \rho & 0 & 0 & 0 \\ 0 & P & 0 & 0 \\ 0 & 0 & P & 0 \\ 0 & 0 & 0 & P \end{bmatrix} \rightarrow \begin{bmatrix} \rho & 0 & 0 & 0 \\ 0 & 0 & 0 & 0 \\ 0 & 0 & 0 & 0 \\ 0 & 0 & 0 & 0 \end{bmatrix}$$

In the above matrix we see that the P is tending to zero because, for Newton's gravity, the mass density is the source of gravity.

The equation of motion is given as:

$$\frac{du^i}{d\tau} + \Gamma^i_{v\alpha} u^v u^\alpha = 0$$

$$\frac{du^i}{d\tau} + \Gamma^{00}_i = 0$$

$$\frac{du^i}{d\tau} + \frac{1}{2}\frac{\partial g_{00}}{\partial x^i} = 0$$

$$\frac{du^i}{d\tau} + \frac{\partial \phi}{\partial x^i} = 0$$

$$g_{00} = -(1 + 2\phi)$$

But we know that $\nabla^2 \phi = 4\pi G\rho$

Therefore,

$$R^{\mu v} = -8\pi G T^{\mu v}$$

Where $-8\pi G T^{\mu v}$ is the constant.

Einstein Tensor

Einstein tensor is also known as trace-reversed Ricci tensor. In Einstein Field Equation, it is used for describing spacetime curvature such that it is in alignment with the conservation of energy and momentum. It is defined as:

$$G = R - \tfrac{1}{2} gR$$

Where,

- R is the Ricci tensor

- g is the metric tensor

- R is the scalar curvature

Stress-energy Tensor

Stress-energy tensor is defined as the tensor $T^{\alpha\beta}$ is a symmetrical tensor which is used for describing the energy and momentum density of the gravitational field. It is given as:

$$T^{\alpha\beta} = T^{\beta\alpha}$$

References

- Topics-relativity-general: physicsoftheuniverse.com, Retrieved 17 July, 2019

- "Even Phenomenally Dense Neutron Stars Fall like a Feather – Einstein Gets It Right Again". Charles Blue, Paul Vosteen. NRAO. 4 July 2018. Retrieved 28 July 2018

- A-Penrose-Diagrams-and-Causality, A-Symmetries, A-General-Relativity-(Crowell), Relativity, Bookshelves: phys.libretexts.org, Retrieved 18 August, 2019

- Hans Christian Von Bayer, The Fermi Solution: Essays on Science, Courier Dover Publications (2001), ISBN 0-486-41707-7, page 79

- Einstein-field-equation, physics: byjus.com, Retrieved 19 January, 2019

- Arnowitt, R.; Deser, S.; Misner, C. (2008). "Republication of: The dynamics of general relativity". General Relativity and Gravitation. 40 (9): 1997–2027. Arxiv:gr-qc/0405109. Bibcode:2008gregr..40.1997A. Doi:10.1007/s10714-008-0661-1

- Kiefer, Claus (2007). Quantum Gravity. Oxford, New York: Oxford University Press. ISBN 978-0-19-921252-1

Phenomena of General Relativity | 5

- **Black Hole**
- **Event Horizon**
- **Two-body Problem in General Relativity**
- **Frame-dragging**
- **Gravitational Singularity**
- **Gravitational Time Dilation**
- **Gravitational Redshift**
- **Shapiro Time Delay**
- **Gravitational Wave**
- **Gravitational Lensing**

Some of the common phenomena which are studied within general relativity are black hole, event horizon, frame-dragging, gravitational singularity, gravitational time dilation, gravitational redshift, Shapiro time delay, gravitational wave and gravitational lensing. This chapter has been carefully written to provide an easy understanding of these phenomena of general relativity.

Black Hole

Black holes are volumes of space where gravity is extreme enough to prevent the escape of even the fastest moving particles. Not even light can break free, hence the name 'black' hole.

A German physicist and astronomer named Karl Schwarzschild proposed the modern version of a black hole in 1915 after coming up with an exact solution to Einstein's approximations of general relativity.

Schwarzschild realised it was possible for mass to be squeezed into an infinitely small point. This would make spacetime around it bend so that nothing – not even massless photons of light – could escape its curvature.

The cusp of the black hole's slide into oblivion is today referred to as its event horizon, and the distance between this boundary and the infinitely dense core – or singularity – is named after Schwarzschild.

Theoretically, all masses have a Schwarzschild radius that can be calculated. If the Sun's mass was squeezed into an infinitely small point, it would form a black hole with a radius of just under 3 kilometres (about 2 miles).

Similarly, Earth's mass would have a Schwarzschild radius of just a few millimetres, making a black hole no bigger than a marble.

For decades, black holes were exotic peculiarities of general relativity. Physicists have became increasingly confident in their existence as other extreme astronomical objects, such as neutron stars, were discovered. Today it's believed most galaxies have monstrous black holes at their core.

Black Holes Formation

It's generally accepted that stars with a mass at least three times greater than that of our Sun's can undergo extreme gravitational collapse once their fuel depletes.

With so much mass in a confined volume, the collective force of gravity overcomes the rule that usually keeps the building blocks of atoms from occupying the same space. All this density creates a black hole.

A second type of miniature black hole has been hypothesised, though never observed. They're thought to have formed when the rippling vacuum of the early Universe rapidly expanded in an event known as inflation, causing highly dense regions to collapse.

Event Horizon

In astrophysics, an event horizon is a boundary beyond which events cannot affect an observer on the opposite side of it. An event horizon is most commonly associated with black holes, where gravitational forces are so strong that light cannot escape.

Any object approaching the horizon from the observer's side appears to slow down and never quite pass through the horizon, with its image becoming more and more redshifted as time elapses. This means that the wavelength of the light emitted from the object is getting longer as the object moves away from the observer. The notion of an event horizon was originally restricted to black holes; light originating inside an event horizon could cross it temporarily but would return. Later, 1958, a strict definition was introduced by David Finkelstein as a boundary beyond which events cannot affect any outside observer at all, encompassing other scenarios than black holes. This strict definition of EH has caused information and firewall paradoxes; therefore Stephen Hawking has supposed an apparent horizon to be used, saying "gravitational collapse produces apparent horizons but no event horizons" and "The absence of event horizons mean that there are no black holes - in the sense of regimes from which light can't escape to infinity"

The black hole event horizon is teleological in nature, meaning that we need to know the entire future space-time of the universe to determine the current location of the horizon, which is essentially impossible. Because of the purely theoretical nature of the event horizon boundary, the traveling object does not necessarily experience strange effects and does, in fact, pass through the calculatory boundary in a finite amount of proper time.

More specific types of horizon include the related but distinct absolute and apparent horizons found around a black hole. Still other distinct notions include the Cauchy and Killing horizons; the photon spheres and ergospheres of the Kerr solution; particle and cosmological horizons relevant to cosmology; and isolated and dynamical horizons important in current black hole research.

Event Horizon of a Black Hole

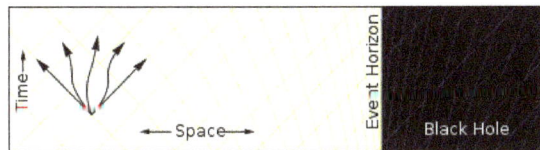

Far away from the black hole a particle can move in any direction. It is only restricted by the speed of light.

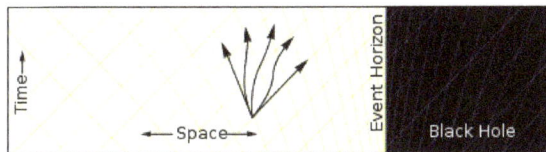

Closer to the black hole spacetime starts to deform. In some convenient coordinate systems, there are more paths going towards the black hole than paths moving away.

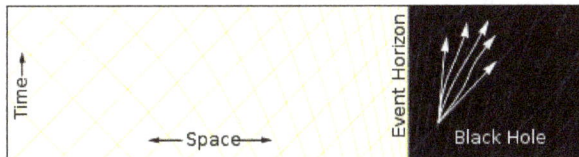

Inside the event horizon all paths bring the particle closer to the center of the black hole. It is no longer possible for the particle to escape.

One of the best-known examples of an event horizon derives from general relativity's description of a black hole, a celestial object so massive that no nearby matter or radiation can escape its gravitational field. Often, this is described as the boundary within which the black hole's escape velocity is greater than the speed of light. However, a more accurate description is that within this horizon, all lightlike paths (paths that light could take) and hence all paths in the forward light cones of particles within the horizon, are warped so as to fall farther into the hole. Once a particle is inside the horizon, moving into the hole is as inevitable as moving forward in time, and can actually be thought of as equivalent to doing so, depending on the spacetime coordinate system used.

The surface at the Schwarzschild radius acts as an event horizon in a non-rotating body that fits inside this radius (although a rotating black hole operates slightly differently). The Schwarzschild radius of an object is proportional to its mass. Theoretically, any amount of matter will become a black hole if compressed into a space that fits within its corresponding Schwarzschild radius. For the mass of the Sun this radius is approximately 3 kilometers and for the Earth it is about 9 millimeters. In practice, however, neither the Earth nor the Sun has the necessary mass and therefore the necessary gravitational force, to overcome electron and neutron degeneracy pressure. The minimal mass required for a star to be able to collapse beyond these pressures is the Tolman–Oppenheimer–Volkoff limit, which is approximately three solar masses.

Black hole event horizons are widely misunderstood. Common, although erroneous, is the notion that black holes "vacuum up" material in their neighborhood, where in fact they are no more capable of seeking out material to consume than any other gravitational attractor. As with any mass in the universe, matter must come within its gravitational scope for the possibility to exist of capture or consolidation with any other mass. Equally common is the idea that matter can be observed falling into a black hole. This is not possible. Astronomers can detect only accretion disks around black holes, where material moves with such speed that friction creates high-energy radiation which can be detected (similarly, some matter from these accretion disks is forced out along the axis of spin of the black hole, creating visible jets when these streams interact with matter such as interstellar gas or when they happen to be aimed directly at Earth). Furthermore, a distant observer will never actually see something reach the horizon. Instead, while approaching the hole, the object will seem to go ever more slowly, while any light it emits will be further and further redshifted.

Cosmic Event Horizon

In cosmology, the event horizon of the observable universe is the largest comoving distance from which light emitted *now* can ever reach the observer in the future. This differs from the concept of particle horizon, which represents the largest comoving distance from which light emitted in the *past* could have reached the observer at a given time. For events beyond that distance, light has not had time to reach our location, even if it were emitted at the time the universe began. How the particle horizon changes with time depends on the nature of the expansion of the universe. If the expansion has certain characteristics, there are parts of the universe that will never be observable, no matter how long the observer waits for light from those regions to arrive. The boundary past which events cannot ever be observed is an event horizon, and it represents the maximum extent of the particle horizon.

The criterion for determining whether a particle horizon for the universe exists is as follows. Define a comoving distance d_p as:

$$d_p = \int_0^{t_0} \frac{c}{a(t)} dt.$$

In this equation, a is the scale factor, c is the speed of light, and t_0 is the age of the Universe. If $d_p \to \infty$ (i.e., points arbitrarily as far away as can be observed), then no event horizon exists. If $d_p \neq \infty$, a horizon is present.

Examples of cosmological models without an event horizon are universes dominated by matter or by radiation. An example of a cosmological model with an event horizon is a universe dominated by the cosmological constant.

A calculation of the speeds of the cosmological event and particle horizons was given in a paper on the FLRW cosmological model, approximating the Universe as composed of non-interacting constituents, each one being a perfect fluid.

Apparent Horizon of an Accelerated Particle

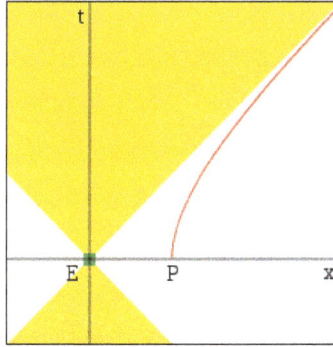

Spacetime diagram showing a uniformly accelerated particle, P, and an event E that is outside
the particle's apparent horizon. The event's forward light cone never intersects the particle's world line.

If a particle is moving at a constant velocity in a non-expanding universe free of gravitational
fields, any event that occurs in that Universe will eventually be observable by the particle, because
the forward light cones from these events intersect the particle's world line. On the other hand,
if the particle is accelerating, in some situations light cones from some events never intersect the
particle's world line. Under these conditions, an apparent horizon is present in the particle's (accelerating) reference frame, representing a boundary beyond which events are unobservable.

For example, this occurs with a uniformly accelerated particle. A spacetime diagram of this situation is shown in the figure. As the particle accelerates, it approaches, but never reaches, the speed
of light with respect to its original reference frame. On the spacetime diagram, its path is a hyperbola, which asymptotically approaches a 45-degree line (the path of a light ray). An event whose
light cone's edge is this asymptote or is farther away than this asymptote can never be observed
by the accelerating particle. In the particle's reference frame, there is a boundary behind it from
which no signals can escape (an apparent horizon). The distance to this boundary is given by c^2 / a
where a is the constant proper acceleration of the particle.

While approximations of this type of situation can occur in the real world (in particle accelerators,
for example), a true event horizon is never present, as this requires the particle to be accelerated
indefinitely (requiring arbitrarily large amounts of energy and an arbitrarily large apparatus).

Interacting with an Event Horizon

A misconception concerning event horizons, especially black hole event horizons, is that they
represent an immutable surface that destroys objects that approach them. In practice, all event
horizons appear to be some distance away from any observer, and objects sent towards an event
horizon never appear to cross it from the sending observer's point of view (as the horizon-crossing
event's light cone never intersects the observer's world line). Attempting to make an object near
the horizon remain stationary with respect to an observer requires applying a force whose magnitude increases unboundedly (becoming infinite) the closer it gets.

In the case of a horizon perceived by a uniformly accelerating observer in empty space, the horizon
seems to remain a fixed distance from the observer no matter how its surroundings move. Varying
the observer's acceleration may cause the horizon to appear to move over time, or may prevent an

event horizon from existing, depending on the acceleration function chosen. The observer never touches the horizon and never passes a location where it appeared to be.

In the case of a horizon perceived by an occupant of a de Sitter universe, the horizon always appears to be a fixed distance away for a non-accelerating observer. It is never contacted, even by an accelerating observer.

In the case of the horizon around a black hole, observers stationary with respect to a distant object will all agree on where the horizon is. While this seems to allow an observer lowered towards the hole on a rope (or rod) to contact the horizon, in practice this cannot be done. The proper distance to the horizon is finite, so the length of rope needed would be finite as well, but if the rope were lowered slowly (so that each point on the rope was approximately at rest in Schwarzschild coordinates), the proper acceleration (G-force) experienced by points on the rope closer and closer to the horizon would approach infinity, so the rope would be torn apart. If the rope is lowered quickly (perhaps even in freefall), then indeed the observer at the bottom of the rope can touch and even cross the event horizon. But once this happens it is impossible to pull the bottom of rope back out of the event horizon, since if the rope is pulled taut, the forces along the rope increase without bound as they approach the event horizon and at some point the rope must break. Furthermore, the break must occur not at the event horizon, but at a point where the second observer can observe it.

Observers crossing a black hole event horizon can calculate the moment they have crossed it, but will not actually see or feel anything special happen at that moment. In terms of visual appearance, observers who fall into the hole perceive the black region constituting the horizon as lying at some apparent distance below them, and never experience crossing this visual horizon. Other objects that had entered the horizon along the same radial path but at an earlier time would appear below the observer but still above the visual position of the horizon, and if they had fallen in recently enough the observer could exchange messages with them before either one was destroyed by the gravitational singularity. Increasing tidal forces (and eventual impact with the hole's singularity) are the only locally noticeable effects. Tidal forces are a function of the mass of the black hole. In realistic stellar black holes, spaghettification occurs early: tidal forces tear materials apart well before the event horizon. However, in supermassive black holes, which are found in centers of galaxies, spaghettification occurs inside the event horizon. A human astronaut would survive the fall through an event horizon only in a black hole with a mass of approximately 10,000 solar masses or greater.

Beyond General Relativity

The description of event horizons given by general relativity is thought to be incomplete. When the conditions under which event horizons occur are modeled using a more comprehensive picture of the way the Universe works, that includes both relativity and quantum mechanics, event horizons are expected to have properties that are different from those predicted using general relativity alone.

At present, it is expected that the primary impact of quantum effects is for event horizons to possess a temperature and so emit radiation. For black holes, this manifests as Hawking radiation, and the larger question of how the black hole possesses a temperature is part of the topic of black hole thermodynamics. For accelerating particles, this manifests as the Unruh effect, which causes space around the particle to appear to be filled with matter and radiation.

According to the controversial black hole firewall hypothesis, matter falling into a black hole would be burned to a crisp by a high energy "firewall" at the event horizon.

An alternative is provided by the complementarity principle, according to which, in the chart of the far observer, infalling matter is thermalized at the horizon and reemitted as Hawking radiation, while in the chart of an infalling observer matter continues undisturbed through the inner region and is destroyed at the singularity. This hypothesis does not violate the no-cloning theorem as there is a single copy of the information according to any given observer. Black hole complementarity is actually suggested by the scaling laws of strings approaching the event horizon, suggesting that in the Schwarzschild chart they stretch to cover the horizon and thermalize into a Planck length-thick membrane.

A complete description of event horizons is expected to, at minimum, require a theory of quantum gravity. One such candidate theory is M-theory. Another such candidate theory is loop quantum gravity.

Two-body Problem in General Relativity

The two-body problem in general relativity is the determination of the motion and gravitational field of two bodies as described by the field equations of general relativity. Solving the Kepler problem is essential to calculate the bending of light by gravity and the motion of a planet orbiting its sun. Solutions are also used to describe the motion of binary stars around each other, and estimate their gradual loss of energy through gravitational radiation. It is customary to assume that both bodies are point-like, so that tidal forces and the specifics of their material composition can be neglected.

General relativity describes the gravitational field by curved space-time; the field equations governing this curvature are nonlinear and therefore difficult to solve in a closed form. No exact solutions of the Kepler problem have been found, but an approximate solution has: the Schwarzschild solution. This solution pertains when the mass M of one body is overwhelmingly greater than the mass m of the other. If so, the larger mass may be taken as stationary and the sole contributor to the gravitational field. This is a good approximation for a photon passing a star and for a planet orbiting its sun. The motion of the lighter body (called the "particle") can then be determined from the Schwarzschild solution; the motion is a geodesic ("shortest path between two points") in the curved space-time. Such geodesic solutions account for the anomalous precession of the planet Mercury, which is a key piece of evidence supporting the theory of general relativity. They also describe the bending of light in a gravitational field, another prediction famously used as evidence for general relativity.

If both masses are considered to contribute to the gravitational field, as in binary stars, the Kepler problem can be solved only approximately. The earliest approximation method to be developed was the post-Newtonian expansion, an iterative method in which an initial solution is gradually corrected. More recently, it has become possible to solve Einstein's field equation using a computer instead of mathematical formulae. As the two bodies orbit each other, they will emit gravitational radiation; this causes them to lose energy and angular momentum gradually, as illustrated by the binary pulsar PSR B1913+16.

For binary black holes numerical solution of the two body problem was achieved after four decades of research, in 2005, when three groups devised the breakthrough techniques.

Classical Kepler Problem

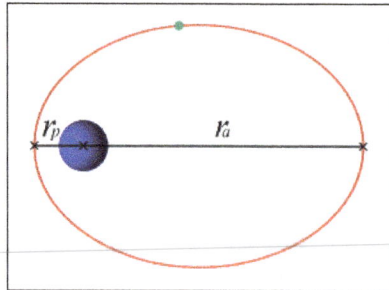

Typical elliptical path of a smaller mass m orbiting a much larger mass M. The larger mass is also moving on an elliptical orbit, but it is too small to be seen because M is much greater than m. The ends of the diameter indicate the apsides, the points of closest and farthest distance.

The Kepler problem derives its name from Johannes Kepler, who worked as an assistant to the Danish astronomer Tycho Brahe. Brahe took extraordinarily accurate measurements of the motion of the planets of the Solar System. From these measurements, Kepler was able to formulate Kepler's laws, the first modern description of planetary motion:

1. The orbit of every planet is an ellipse with the Sun at one of the two foci.

2. A line joining a planet and the Sun sweeps out equal areas during equal intervals of time.

3. The square of the orbital period of a planet is directly proportional to the cube of the semi-major axis of its orbit.

Kepler published the first two laws in 1609 and the third law in 1619. They supplanted earlier models of the Solar System, such as those of Ptolemy and Copernicus. Kepler's laws apply only in the limited case of the two-body problem. Voltaire and Émilie du Châtelet were the first to call them "Kepler's laws".

Nearly a century later, Isaac Newton had formulated his three laws of motion. In particular, Newton's second law states that a force F applied to a mass m produces an acceleration a given by the equation $F=ma$. Newton then posed the question: what must the force be that produces the elliptical orbits seen by Kepler? His answer came in his law of universal gravitation, which states that the force between a mass M and another mass m is given by the formula:

$$F = G\frac{Mm}{r^2}.$$

where, r is the distance between the masses and G is the gravitational constant. Given this force law and his equations of motion, Newton was able to show that two point masses attracting each other would each follow perfectly elliptical orbits. The ratio of sizes of these ellipses is m/M, with the larger mass moving on a smaller ellipse. If M is much larger than m, then the larger mass will appear to be stationary at the focus of the elliptical orbit of the lighter mass m. This model can be applied approximately to the Solar System. Since the mass of the Sun is much larger than those of the planets, the force acting on each planet is principally due to the Sun; the gravity of the planets for each other can be neglected to first approximation.

Apsidal Precession

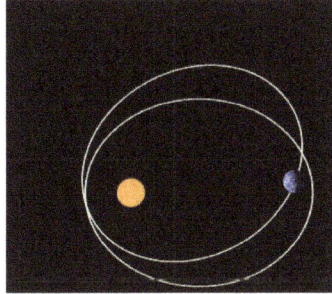

In the absence of any other forces, a particle orbiting another under the influence of Newtonian gravity follows the same perfect ellipse eternally. The presence of other forces (such as the gravitation of other planets), causes this ellipse to rotate gradually. The rate of this rotation (called orbital precession) can be measured very accurately. The rate can also be predicted knowing the magnitudes and directions of the other forces. However, the predictions of Newtonian gravity do not match the observations, as discovered in 1859 from observations of Mercury.

If the potential energy between the two bodies is not exactly the $1/r$ potential of Newton's gravitational law but differs only slightly, then the ellipse of the orbit gradually rotates (among other possible effects). This apsidal precession is observed for all the planets orbiting the Sun, primarily due to the oblateness of the Sun (it is not perfectly spherical) and the attractions of the other planets to one another. The apsides are the two points of closest and furthest distance of the orbit (the periapsis and apoapsis, respectively); apsidal precession corresponds to the rotation of the line joining the apsides. It also corresponds to the rotation of the Laplace–Runge–Lenz vector, which points along the line of apsides.

Newton's law of gravitation soon became accepted because it gave very accurate predictions of the motion of all the planets. These calculations were carried out initially by Pierre-Simon Laplace in the late 18th century, and refined by Félix Tisserand in the later 19th century. Conversely, if Newton's law of gravitation did *not* predict the apsidal precessions of the planets accurately, it would have to be discarded as a theory of gravitation. Such an anomalous precession was observed in the second half of the 19th century.

Anomalous Precession of Mercury

In 1859, Urbain Le Verrier discovered that the orbital precession of the planet Mercury was not quite what it should be; the ellipse of its orbit was rotating (precessing) slightly faster than predicted by the traditional theory of Newtonian gravity, even after all the effects of the other planets had been accounted for. The effect is small (roughly 43 arcseconds of rotation per century), but well above the measurement error (roughly 0.1 arcseconds per century). Le Verrier realized the importance of his discovery immediately, and challenged astronomers and physicists alike to account for it. Several classical explanations were proposed, such as interplanetary dust, unobserved oblateness of the Sun, an undetected moon of Mercury, or a new planet named Vulcan. After these explanations were discounted, some physicists were driven to the more radical hypothesis that Newton's inverse-square law of gravitation was incorrect. For example, some physicists proposed a power law with an exponent that was slightly different from 2.

Others argued that Newton's law should be supplemented with a velocity-dependent potential. However, this implied a conflict with Newtonian celestial dynamics. In his treatise on celestial mechanics, Laplace had shown that if the gravitational influence does not act instantaneously, then the motions of the planets themselves will not exactly conserve momentum (and consequently some of the momentum would have to be ascribed to the mediator of the gravitational interaction, analogous to ascribing momentum to the mediator of the electromagnetic interaction.) As seen from a Newtonian point of view, if gravitational influence does propagate at a finite speed, then at all points in time a planet is attracted to a point where the Sun was some time before, and not towards the instantaneous position of the Sun. On the assumption of the classical fundamentals, Laplace had shown that if gravity would propagate at a velocity on the order of the speed of light then the solar system would be unstable, and would not exist for a long time. The observation that the solar system is old enough allowed him to put a lower limit on the speed of gravity that turned out to be many orders of magnitude faster than the speed of light.

Laplace's estimate for the velocity of gravity is not correct in a field theory which respects the principle of relativity. Since electric and magnetic fields combine, the attraction of a point charge which is moving at a constant velocity is towards the extrapolated instantaneous position, not to the apparent position it seems to occupy when looked at. To avoid those problems, between 1870 and 1900 many scientists used the electrodynamic laws of Wilhelm Eduard Weber, Carl Friedrich Gauss, Bernhard Riemann to produce stable orbits and to explain the perihelion shift of Mercury's orbit. In 1890 Lévy succeeded in doing so by combining the laws of Weber and Riemann, whereby the speed of gravity is equal to the speed of light in his theory. And in another attempt Paul Gerber even succeeded in deriving the correct formula for the perihelion shift (which was identical to that formula later used by Einstein). However, because the basic laws of Weber and others were wrong (for example, Weber's law was superseded by Maxwell's theory), those hypotheses were rejected. Another attempt by Hendrik Lorentz, who already used Maxwell's theory, produced a perihelion shift which was too low.

Einstein's Theory of General Relativity

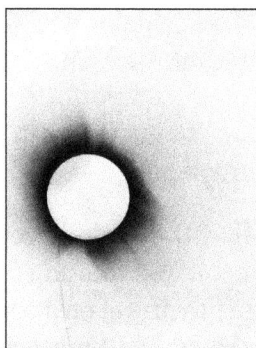

Eddington's 1919 measurements of the bending of star-light by the Sun's gravity led to the acceptance of general relativity worldwide.

Around 1904–1905, the works of Hendrik Lorentz, Henri Poincaré and finally Albert Einstein's special theory of relativity, exclude the possibility of propagation of any effects faster than the speed of light. It followed that Newton's law of gravitation would have to be replaced with another law, compatible with the principle of relativity, while still obtaining the newtonian limit for circumstances where relativistic effects are negligible. Such attempts were made by Henri Poincaré, Hermann Minkowski and Arnold Sommerfeld. In 1907 Einstein came to the conclusion that to achieve this a successor to special relativity was needed. From 1907 to 1915, Einstein worked towards a new theory, using his equivalence principle as a key concept to guide his way. According to this principle, a uniform gravitational field acts equally on everything within it and, therefore, cannot be detected by a free-falling observer. Conversely, all local gravitational effects should be reproducible in a linearly accelerating reference frame, and vice versa. Thus, gravity acts like a fictitious force such as the centrifugal force or the Coriolis force, which result from being in an accelerated reference frame; all fictitious forces are proportional to the inertial mass, just as gravity is. To effect the reconciliation of gravity and special relativity and to incorporate the equivalence principle, something had to be sacrificed; that something was the long-held classical assumption that our space obeys the laws of Euclidean geometry, e.g., that the Pythagorean theorem is true experimentally. Einstein used a more general geometry, pseudo-Riemannian geometry, to allow for the curvature of space and time that was necessary for the reconciliation; after eight years of work, he succeeded in discovering the precise way in which space-time should be curved in order to reproduce the physical laws observed in Nature, particularly gravitation. Gravity is distinct from the fictitious forces centrifugal force and coriolis force in the sense that the curvature of spacetime is regarded as physically real, whereas the fictitious forces are not regarded as forces. The very first solutions of his field equations explained the anomalous precession of Mercury and predicted an unusual bending of light, which was confirmed *after* his theory was published. These solutions are explained.

General Relativity, Special Relativity and Geometry

In the normal Euclidean geometry, triangles obey the Pythagorean theorem, which states that the square distance ds2 between two points in space is the sum of the squares of its perpendicular components:

$$ds^2 = dx^2 + dy^2 + dz^2$$

where dx, dy and dz represent the infinitesimal differences between the x, y and z coordinates of two points in a Cartesian coordinate system . Now imagine a world in which this is not quite true; a world where the distance is instead given by:

$$ds^2 = F(x,y,z)dx^2 + G(x,y,z)dy^2 + H(x,y,z)dz^2$$

where F, G and H are arbitrary functions of position. It is not hard to imagine such a world; we live on one. The surface of the earth is curved, which is why it's impossible to make a perfectly accurate flat map of the earth. Non-Cartesian coordinate systems illustrate this well; for example, in the spherical coordinates (r, θ, φ), the Euclidean distance can be written:

$$ds^2 = dr^2 + r^2 d\theta^2 + r^2 \sin^2 \theta d\varphi^2$$

Another illustration would be a world in which the rulers used to measure length were untrustworthy, rulers that changed their length with their position and even their orientation. In the most general case, one must allow for cross-terms when calculating the distance ds:

$$ds^2 = g_{xx}dx^2 + g_{xy}dxdy + g_{xz}dxdz + \cdots + g_{zy}dzdy + g_{zz}dz^2$$

where the nine functions $g_{xx}, g_{xy}, \ldots, g_{zz}$ constitute the metric tensor, which defines the geometry of the space in Riemannian geometry. In the spherical-coordinates example above, there are no cross-terms; the only nonzero metric tensor components are $g_{rr} = 1$, $g_{\theta\theta} = r^2$ and $g_{\varphi\varphi} = r^2 \sin^2 \theta$.

In his special theory of relativity, Albert Einstein showed that the distance ds between two spatial points is not constant, but depends on the motion of the observer. However, there is a measure of separation between two points in space-time — called "proper time" and denoted with the symbol $d\tau$ — that *is* invariant; in other words, it doesn't depend on the motion of the observer.

$$c^2d\tau^2 = c^2dt^2 - dx^2 - dy^2 - dz^2$$

which may be written in spherical coordinates as:

$$c^2d\tau^2 = c^2dt^2 - dr^2 - r^2d\theta^2 - r^2 \sin^2 \theta d\varphi^2$$

This formula is the natural extension of the Pythagorean theorem and similarly holds only when there is no curvature in space-time. In general relativity, however, space and time may have curvature, so this distance formula must be modified to a more general form:

$$c^2d\tau^2 = g_{\mu\nu}dx^\mu dx^\nu$$

just as we generalized the formula to measure distance on the surface of the Earth. The exact form of the metric $g\mu v$ depends on the gravitating mass, momentum and energy, as described by the Einstein field equations. Einstein developed those field equations to match the then known laws of Nature; however, they predicted never-before-seen phenomena (such as the bending of light by gravity) that were confirmed later.

Geodesic Equation

According to Einstein's theory of general relativity, particles of negligible mass travel along geodesics in the space-time. In uncurved space-time, far from a source of gravity, these geodesics correspond to straight lines; however, they may deviate from straight lines when the space-time is curved. The equation for the geodesic lines is:

$$\frac{d^2x^\mu}{dq^2} + \Gamma^\mu_{\nu\lambda}\frac{dx^\nu}{dq}\frac{dx^\lambda}{dq} = 0$$

where Γ represents the Christoffel symbol and the variable q parametrizes the particle's path through space-time, its so-called world line. The Christoffel symbol depends only on the metric tensor $g_{\mu\nu}$, or rather on how it changes with position. The variable q is a constant multiple of the proper time τ for timelike orbits (which are traveled by massive particles), and is usually taken to be equal to it. For lightlike (or null) orbits (which are traveled by massless particles such as the photon), the proper time is zero and, strictly speaking, cannot be used as the variable q. Nevertheless, lightlike orbits can be derived as the ultrarelativistic limit of timelike orbits, that is, the limit as the particle mass m goes to zero while holding its total energy fixed.

Schwarzschild Solution

An exact solution to the Einstein field equations is the Schwarzschild metric, which corresponds to the external gravitational field of a stationary, uncharged, non-rotating, spherically symmetric body of mass M. It is characterized by a length scale r_s, known as the Schwarzschild radius, which is defined by the formula:

$$r_s = \frac{2GM}{c^2}$$

where G is the gravitational constant. The classical Newtonian theory of gravity is recovered in the limit as the ratio r_s/r goes to zero. In that limit, the metric returns to that defined by special relativity.

In practice, this ratio is almost always extremely small. For example, the Schwarzschild radius r_s of the Earth is roughly 9 mm ($\frac{3}{8}$ inch); at the surface of the Earth, the corrections to Newtonian gravity are only one part in a billion. The Schwarzschild radius of the Sun is much larger, roughly 2953 meters, but at its surface, the ratio r_s/r is roughly 4 parts in a million. A white dwarf star is much denser, but even here the ratio at its surface is roughly 250 parts in a million. The ratio only becomes large close to ultra-dense objects such as neutron stars (where the ratio is roughly 50%) and black holes.

Orbits about the Central Mass

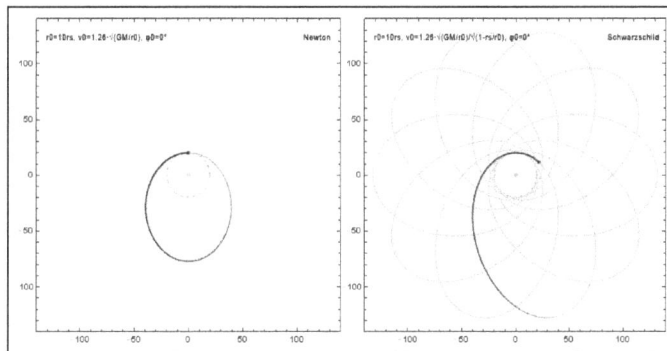

Comparison between the orbit of a testparticle in Newtonian (left) and Schwarzschild (right) spacetime.

The orbits of a test particle of infinitesimal mass m about the central mass M is given by the equation of motion:

$$\left(\frac{dr}{d}\right) = \frac{E^2}{m^2c^2} - \left(1 - \frac{}{r}\right)\left(+ \frac{h^2}{r^2}\right).$$

Where h is the specific relative angular momentum, $h = r \times v = \dfrac{L}{\mu}$ and μ is the reduced mass. This can be converted into an equation for the orbit:

$$\left(\frac{dr}{d\varphi}\right)^2 = \frac{r^4}{b^2} - \left(1 - \frac{r_s}{r}\right)\left(\frac{r^4}{a^2} + r^2\right)$$

where, for brevity, two length-scales, $a = h/c$ and $b = Lc/E$, have been introduced. They are constants of the motion and depend on the initial conditions (position and velocity) of the test particle. Hence, the solution of the orbit equation is:

$$\varphi = \int \frac{1}{r^2}\left[\frac{1}{b^2} - \left(1 - \frac{r_s}{r}\right)\left(\frac{1}{a^2} + \frac{1}{r^2}\right)\right]^{-1/2} dr.$$

Effective Radial Potential Energy

The equation of motion for the particle derived above:

$$\left(\frac{dr}{d\tau}\right)^2 = \frac{E^2}{m^2c^2} - c^2 + \frac{r_s c^2}{r} - \frac{h^2}{r^2} + \frac{r_s h^2}{r^3}$$

can be rewritten using the definition of the Schwarzschild radius r_s as:

$$\frac{1}{2} m \left(\frac{dr}{d\tau}\right)^2 = \left[\frac{E^2}{2mc^2} - \frac{1}{2}mc^2\right] + \frac{GMm}{r} - \frac{L^2}{2\mu r^2} + \frac{G(M+m)L^2}{c^2 \mu r^3}$$

which is equivalent to a particle moving in a one-dimensional effective potential:

$$V(r) = -\frac{GMm}{r} + \frac{L^2}{2\mu r^2} - \frac{G(M+m)L^2}{c^2 \mu r^3}$$

The first two terms are well-known classical energies, the first being the attractive Newtonian gravitational potential energy and the second corresponding to the repulsive "centrifugal" potential energy; however, the third term is an attractive energy unique to general relativity. As shown, this inverse-cubic energy causes elliptical orbits to precess gradually by an angle $\delta\varphi$ per revolution:

$$\delta\varphi \approx \frac{6\pi G(M+m)}{c^2 A\left(1 - e^2\right)}$$

where A is the semi-major axis and e is the eccentricity. Here $\delta\varphi$ is NOT the change in the φ-coordinate in (t, r, θ, φ) coordinates but the change in the argument of periapsis of the classical closed orbit.

The third term is attractive and dominates at small r values, giving a critical inner radius r_{inner}

at which a particle is drawn inexorably inwards to $r=0$; this inner radius is a function of the particle's angular momentum per unit mass or, equivalently, the a length-scale defined above.

Circular Orbits and their Stability

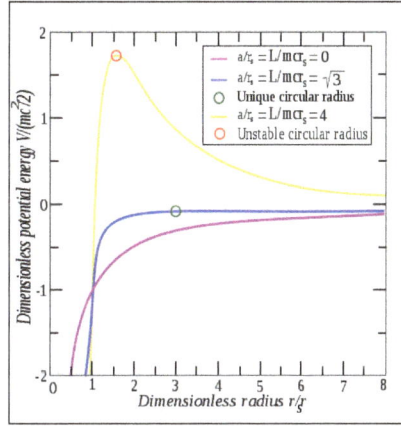

Effective radial potential for various angular momenta. At small radii, the energy drops precipitously, causing the particle to be pulled inexorably inwards to $r=0$. However, when the normalized angular momentum $a/r_s = L/mcr_s$ equals the square root of three, a metastable circular orbit is possible at the radius highlighted with a green circle. At higher angular momenta, there is a significant centrifugal barrier (orange curve) and an unstable inner radius, highlighted in red.

The effective potential V can be re-written in terms of the length $a = h/c$:

$$V(r) = \frac{mc^2}{2}\left[-\frac{r_s}{r} + \frac{a^2}{r^2} - \frac{r_s a^2}{r^3}\right].$$

Circular orbits are possible when the effective force is zero:

$$F = -\frac{dV}{dr} = -\frac{mc^2}{2r^4}\left[r_s r^2 - 2a^2 r + 3r_s a^2\right] = 0;$$

i.e., when the two attractive forces—Newtonian gravity (first term) and the attraction unique to general relativity (third term)—are exactly balanced by the repulsive centrifugal force (second term). There are two radii at which this balancing can occur, denoted here as r_{inner} and r_{outer}:

$$r_{outer} = \frac{a^2}{r_s}\left(1 + \sqrt{1 - \frac{3r_s^2}{a^2}}\right)$$

$$r_{inner} = \frac{a^2}{r_s}\left(1 - \sqrt{1 - \frac{3r_s^2}{a^2}}\right) = \frac{3a^2}{r_{outer}},$$

which are obtained using the quadratic formula. The inner radius rinner is unstable, because the attractive third force strengthens much faster than the other two forces when r becomes small; if the particle slips slightly inwards from rinner (where all three forces are in balance), the third force dominates the other two and draws the particle inexorably inwards to r = o. At the outer radius, however, the circular orbits are stable; the third term is less important and the system behaves more like the non-relativistic Kepler problem.

When a is much greater than r_s (the classical case), these formulae become approximately:

$$r_{outer} \approx \frac{2a^2}{r_s}$$

$$r_{inner} \approx \frac{3}{2}r_s$$

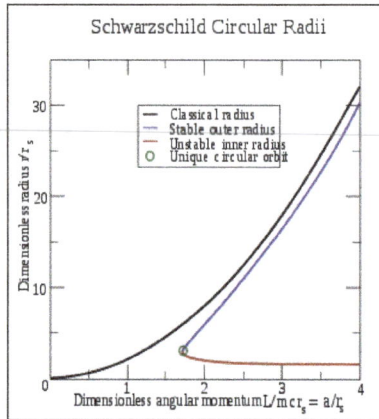

The stable and unstable radii are plotted versus the normalized angular momentum $a/r_s = L/mcr_s$ in blue and red, respectively. These curves meet at a unique circular orbit (green circle) when the normalized angular momentum equals the square root of three. For comparison, the classical radius predicted from the centripetal acceleration and Newton's law of gravity is plotted in black.

Substituting the definitions of a and r_s into r_{outer} yields the classical formula for a particle of mass m orbiting a body of mass M.

The following equation:

$$r_{outer}^3 = \frac{G(M+m)}{\omega_\varphi^2}$$

where ω_φ is the orbital angular speed of the particle, is obtained in non-relativistic mechanics by setting the centrifugal force equal to the Newtonian gravitational force:

$$\frac{GMm}{r^2} = \mu\omega_\varphi^2 r$$

where μ is the reduced mass.

In our notation, the classical orbital angular speed equals:

$$\omega_\varphi^2 \approx \frac{GM}{r_{outer}^3} = \left(\frac{r_s c^2}{2r_{outer}^3}\right) = \left(\frac{r_s c^2}{2}\right)\left(\frac{r_s^3}{8a^6}\right) = \frac{c^2 r_s^4}{16a^6}$$

At the other extreme, when a^2 approaches $3r_s^2$ from above, the two radii converge to a single value:

$$r_{outer} \approx r_{inner} \approx 3r_s$$

The quadratic solutions above ensure that r_{outer} is always greater than $3r_s$, whereas r_{inner} lies between $\frac{3}{2}r_s$ and $3r_s$. Circular orbits smaller than $\frac{3}{2}r_s$ are not possible. For massless particles, a goes to infinity, implying that there is a circular orbit for photons at $r_{inner} = \frac{3}{2}r_s$. The sphere of this radius is sometimes known as the photon sphere.

Precession of Elliptical Orbits

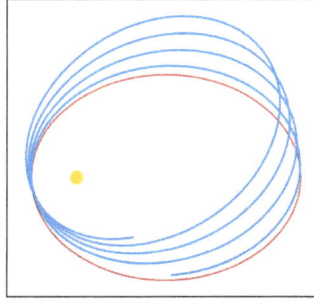

In the non-relativistic Kepler problem, a particle follows the same perfect ellipse (red orbit) eternally. General relativity introduces a third force that attracts the particle slightly more strongly than Newtonian gravity, especially at small radii. This third force causes the particle's elliptical orbit to precess (cyan orbit) in the direction of its rotation; this effect has been measured in Mercury, Venus and Earth. The yellow dot within the orbits represents the center of attraction, such as the Sun.

The orbital precession rate may be derived using this radial effective potential V. A small radial deviation from a circular orbit of radius r_{outer} will oscillate in a stable manner with an angular frequency:

$$\omega_r^2 = \frac{1}{m}\left[\frac{d^2V}{dr^2}\right]_{r=r_{outer}}$$

which equals:

$$\omega_r^2 = \left(\frac{c^2 r_s}{2r_{outer}^4}\right)(r_{outer} - r_{inner}) = \omega_\varphi^2\sqrt{1 - \frac{3r_s^2}{a^2}}$$

Taking the square root of both sides and expanding using the binomial theorem yields the formula:

$$\omega_r = \omega_\varphi\left(1 - \frac{3r_s^2}{4a^2} + \cdots\right)$$

Multiplying by the period T of one revolution gives the precession of the orbit per revolution:

$$\delta\varphi = T\left(\omega_\varphi - \omega_r\right) \approx 2\pi\left(\frac{3r_s^2}{4a^2}\right) = \frac{3\pi m^2 c^2}{2L^2}r_s^2$$

where we have used $\omega_\varphi T = 2n$ and the definition of the length-scale a. Substituting the definition of the Schwarzschild radius r_s gives:

$$\delta\varphi \approx \frac{3\pi m^2 c^2}{2L^2}\left(\frac{4G^2 M^2}{c^4}\right) = \frac{6\pi G^2 M^2 m^2}{c^2 L^2}$$

This may be simplified using the elliptical orbit's semi-major axis A and eccentricity e related by the formula:

$$\frac{h^2}{G(M+m)} = A\left(1-e^2\right)$$

to give the precession angle:

$$\delta\varphi \approx \frac{6\pi G(M+m)}{c^2 A\left(1-e^2\right)}$$

Since the closed classical orbit is an ellipse in general, the quantity $A(1 - e^2)$ is the semi-latus rectum l of the ellipse.

Hence, the final formula of angular apsidal precession for a unit complete revolution is:

$$\delta\varphi \approx \frac{6\pi G(M+m)}{c^2 l}$$

Beyond the Schwarzschild Solution

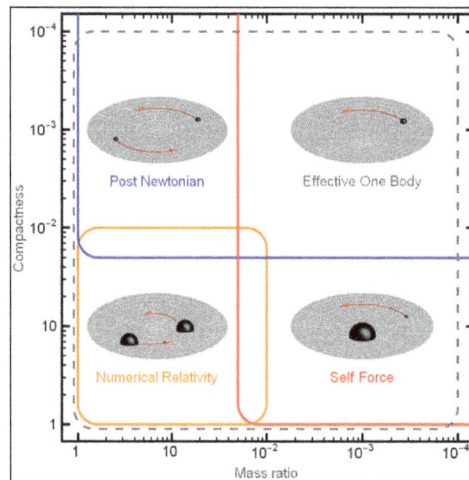

Diagram of the parameter space of compact binaries with the various approximation schemes and their regions of validity.

Post-Newtonian Expansion

In the Schwarzschild solution, it is assumed that the larger mass M is stationary and it alone determines the gravitational field (i.e., the geometry of space-time) and, hence, the lesser mass m

follows a geodesic path through that fixed space-time. This is a reasonable approximation for pho-tons and the orbit of Mercury, which is roughly 6 million times lighter than the Sun. However, it is inadequate for binary stars, in which the masses may be of similar magnitude.

The metric for the case of two comparable masses cannot be solved in closed form and therefore one has to resort to approximation techniques such as the post-Newtonian approximation or nu-merical approximations. In passing, we mention one particular exception in lower dimensions. In (1 + 1) dimensions, i.e. a space made of one spatial dimension and one time dimension, the metric for two bodies of equal masses can be solved analytically in terms of the Lambert W function. However, the gravitational energy between the two bodies is exchanged via dilatons rather than gravitons which require three-space in which to propagate.

The post-Newtonian expansion is a calculational method that provides a series of ever more accu-rate solutions to a given problem. The method is iterative; an initial solution for particle motions is used to calculate the gravitational fields; from these derived fields, new particle motions can be calculated, from which even more accurate estimates of the fields can be computed, and so on. This approach is called "post-Newtonian" because the Newtonian solution for the particle orbits is often used as the initial solution.

When this method is applied to the two-body problem without restriction on their masses, the re-sult is remarkably simple. To the lowest order, the relative motion of the two particles is equivalent to the motion of an infinitesimal particle in the field of their combined masses. In other words, the Schwarzschild solution can be applied, provided that the $M + m$ is used in place of M in the formu-lae for the Schwarzschild radius r_s and the precession angle per revolution $\delta\varphi$.

Modern Computational Approaches

Einstein's equations can also be solved on a computer using sophisticated numerical methods. Given sufficient computer power, such solutions can be more accurate than post-Newtonian solu-tions. However, such calculations are demanding because the equations must generally be solved in a four-dimensional space. Nevertheless, beginning in the late 1990s, it became possible to solve difficult problems such as the merger of two black holes, which is a very difficult version of the Kepler problem in general relativity.

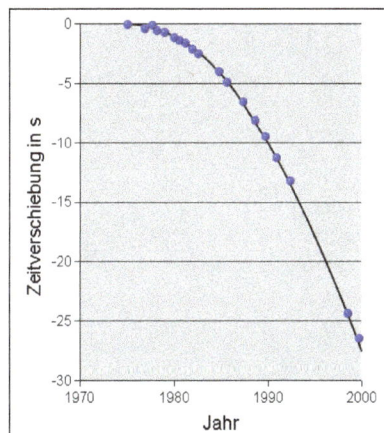

Experimentally observed decreases of the orbital period of the binary pulsar PSR
B1913+16(blue dots) match the predictions of general relativity (black curve) almost exactly.

Gravitational Radiation

If there is no incoming gravitational radiation, according to general relativity, two bodies orbiting one another will emit gravitational radiation, causing the orbits to gradually lose energy.

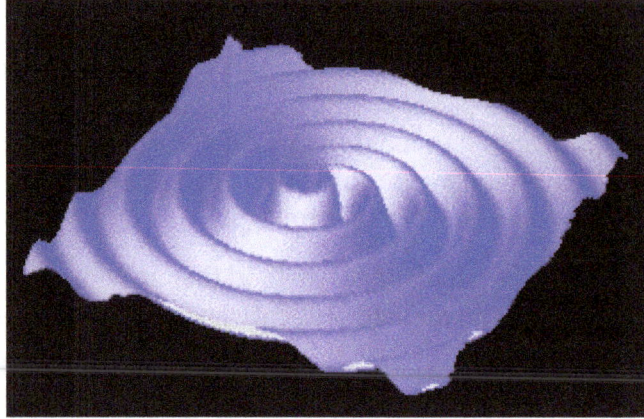

Two neutron stars rotating rapidly around one another gradually lose energy by emitting gravitational radiation. As they lose energy, they orbit each other more quickly and more closely to one another.

The formulae describing the loss of energy and angular momentum due to gravitational radiation from the two bodies of the Kepler problem have been calculated. The rate of losing energy (averaged over a complete orbit) is given by:

$$-\left\langle \frac{dE}{dt} \right\rangle = \frac{32G^4 m_1^2 m_2^2 \left(m_1 + m_2\right)}{5c^5 a^5 \left(1 - e^2\right)^{7/2}} \left(1 + \frac{73}{24}e^2 + \frac{37}{96}e^4\right)$$

where e is the orbital eccentricity and a is the semimajor axis of the elliptical orbit. The angular brackets on the left-hand side of the equation represent the averaging over a single orbit. Similarly, the average rate of losing angular momentum equals:

$$-\left\langle \frac{dL_z}{dt} \right\rangle = \frac{32G^{7/2} m_1^2 m_2^2 \sqrt{m_1 + m_2}}{5c^5 a^{7/2} \left(1 - e^2\right)^2} \left(1 + \frac{7}{8}e^2\right)$$

The rate of period decrease is given by:

$$-\left\langle \frac{dP_b}{dt} \right\rangle = \frac{192\pi G^{5/3} m_1 m_2 \left(m_1 + m_2\right)^{-1/3}}{5c^5 \left(1 - e^2\right)^{7/2}} \left(1 + \frac{73}{24}e^2 + \frac{37}{96}e^4\right) \left(\frac{P_b}{2\pi}\right)^{-5/3}$$

where P_b is orbital period.

The losses in energy and angular momentum increase significantly as the eccentricity approaches one, i.e., as the ellipse of the orbit becomes ever more elongated. The radiation losses also increase significantly with a decreasing size a of the orbit.

Frame-dragging

Frame-dragging is an effect on spacetime, predicted by Einstein's general theory of relativity, that is due to non-static stationary distributions of mass–energy. A stationary field is one that is in a steady state, but the masses causing that field may be non-static, rotating for instance. More generally, the subject of effects caused by mass–energy currents is known as gravitomagnetism, in analogy with classical electromagnetism.

The first frame-dragging effect was derived in 1918, in the framework of general relativity, by the Austrian physicists Josef Lense and Hans Thirring, and is also known as the Lense–Thirring effect. They predicted that the rotation of a massive object would distort the spacetime metric, making the orbit of a nearby test particle precess. This does not happen in Newtonian mechanics for which the gravitational field of a body depends only on its mass, not on its rotation. The Lense–Thirring effect is very small—about one part in a few trillion. To detect it, it is necessary to examine a very massive object, or build an instrument that is very sensitive.

In 2015, new general-relativistic extensions of Newtonian rotation laws were formulated to describe geometric dragging of frames which incorporates a newly discovered antidragging effect.

Effects

Rotational frame-dragging (the Lense–Thirring effect) appears in the general principle of relativity and similar theories in the vicinity of rotating massive objects. Under the Lense–Thirring effect, the frame of reference in which a clock ticks the fastest is one which is revolving around the object as viewed by a distant observer. This also means that light traveling in the direction of rotation of the object will move past the massive object faster than light moving against the rotation, as seen by a distant observer. It is now the best known frame-dragging effect, partly thanks to the Gravity Probe B experiment. Qualitatively, frame-dragging can be viewed as the gravitational analog of electromagnetic induction.

Also, an inner region is dragged more than an outer region. This produces interesting locally rotating frames. For example, imagine that a north-south–oriented ice skater, in orbit over the equator of a black hole and rotationally at rest with respect to the stars, extends her arms. The arm extended toward the black hole will be "torqued" spinward due to gravitomagnetic induction ("torqued" is in quotes because gravitational effects are not considered "forces" under GR). Likewise the arm extended away from the black hole will be torqued anti-spinward. She will therefore be rotationally sped up, in a counter-rotating sense to the black hole. This is the opposite of what happens in everyday experience. There exists a particular rotation rate that, should she be initially rotating at that rate when she extends her arms, inertial effects and frame-dragging effects will balance and her rate of rotation will not change. Due to the equivalence principle, gravitational effects are locally indistinguishable from inertial effects, so this rotation rate, at which when she extends her arms nothing happens, is her local reference for non-rotation. This frame is rotating with respect to the fixed stars and counter-rotating with respect to the black hole. This effect is analogous to the hyperfine structure in atomic spectra due to nuclear spin. A useful metaphor is a planetary gear system with the black hole being the sun gear, the ice skater being a planetary gear and the outside universe being the ring gear.

Another interesting consequence is that, for an object constrained in an equatorial orbit, but not in freefall, it weighs more if orbiting anti-spinward, and less if orbiting spinward. For example, in a suspended equatorial bowling alley, a bowling ball rolled anti-spinward would weigh more than the same ball rolled in a spinward direction. Note, frame dragging will neither accelerate nor slow down the bowling ball in either direction. It is not a "viscosity". Similarly, a stationary plumb-bob suspended over the rotating object will not list. It will hang vertically. If it starts to fall, induction will push it in the spinward direction.

Linear frame dragging is the similarly inevitable result of the general principle of relativity, applied to linear momentum. Although it arguably has equal theoretical legitimacy to the "rotational" effect, the difficulty of obtaining an experimental verification of the effect means that it receives much less discussion and is often omitted from articles on frame-dragging.

Static mass increase is a third effect noted by Einstein in the same paper. The effect is an increase in inertia of a body when other masses are placed nearby. While not strictly a frame dragging effect (the term frame dragging is not used by Einstein), it is demonstrated by Einstein that it derives from the same equation of general relativity. It is also a tiny effect that is difficult to confirm experimentally.

Experimental Tests

In 1976 Van Patten and Everitt proposed to implement a dedicated mission aimed to measure the Lense–Thirring node precession of a pair of counter-orbiting spacecraft to be placed in terrestrial polar orbits with drag-free apparatus. A somewhat equivalent, cheaper version of such an idea was put forth in 1986 by Ciufolini who proposed to launch a passive, geodetic satellite in an orbit identical to that of the LAGEOS satellite, launched in 1976, apart from the orbital planes which should have been displaced by 180 deg apart: the so-called butterfly configuration. The measurable quantity was, in this case, the sum of the nodes of LAGEOS and of the new spacecraft, later named LAGEOS III, LARES, WEBER-SAT.

Limiting the scope to the scenarios involving existing orbiting bodies, the first proposal to use the LAGEOS satellite and the Satellite Laser Ranging (SLR) technique to measure the Lense–Thirring effect dates back to 1977–1978. Tests have started to be effectively performed by using the LAGEOS and LAGEOS II satellites in 1996, according to a strategy involving the use of a suitable combination of the nodes of both satellites and the perigee of LAGEOS II. The latest tests with the LAGEOS satellites have been performed in 2004–2006 by discarding the perigee of LAGEOS II and using a linear combination. Recently, a comprehensive overview of the attempts to measure the Lense-Thirring effect with artificial satellites was published in the literature. The overall accuracy reached in the tests with the LAGEOS satellites is subject to some controversy.

The Gravity Probe B experiment was a satellite-based mission by a Stanford group and NASA, used to experimentally measure another gravitomagnetic effect, the Schiff precession of a gyroscope, to an expected 1% accuracy or better. Unfortunately such accuracy was not achieved. The first preliminary results released in April 2007 pointed towards an accuracy of 256–128%, with the hope of reaching about 13% in December 2007. In 2008 the Senior Review Report of the NASA Astrophysics Division Operating Missions stated that it was unlikely that Gravity Probe B team will be able to reduce the errors to the level necessary to produce a convincing test of currently untested aspects of General Relativity (including frame-dragging). On May 4, 2011, the Stanford-based analysis group and NASA announced the final report, and in it the data from GP-B demonstrated the frame-dragging effect with an error of about 19 percent, and Einstein's predicted value was at the center of the confidence interval.

In the case of stars orbiting close to a spinning, supermassive black hole, frame dragging should cause the star's orbital plane to precess about the black hole spin axis. This effect should be detectable within the next few years via astrometric monitoring of stars at the center of the Milky Way galaxy. By comparing the rate of orbital precession of two stars on different orbits, it is possible in principle to test the no-hair theorems of general relativity, in addition to measuring the spin of the black hole.

Astronomical Evidence

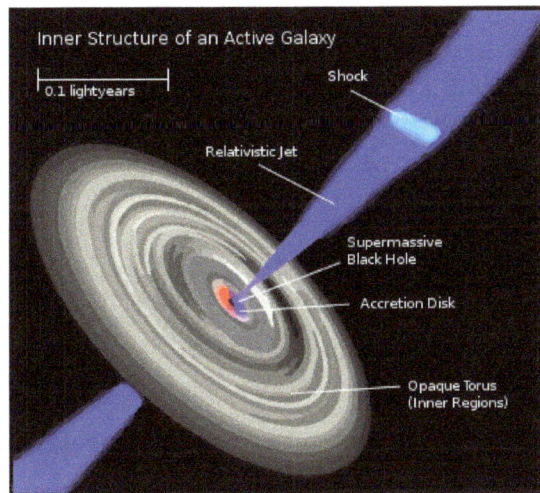

Relativistic jet. The environment around the AGN where the relativistic plasma
is collimated into jets which escape along the pole of the supermassive black hole.

Relativistic jets may provide evidence for the reality of frame-dragging. Gravitomagnetic forces produced by the Lense–Thirring effect (frame dragging) within the ergosphere of rotating black holes combined with the energy extraction mechanism by Penrose have been used to explain the observed properties of relativistic jets. The gravitomagnetic model developed by Reva Kay Williams predicts the observed high energy particles (~GeV) emitted by quasars and active galactic nuclei; the extraction of X-rays, γ-rays, and relativistic e⁻–e⁺ pairs; the collimated jets about the polar axis; and the asymmetrical formation of jets (relative to the orbital plane).

Mathematical Derivation

Frame-dragging may be illustrated most readily using the Kerr metric, which describes the geometry of spacetime in the vicinity of a mass M rotating with angular momentum J, and Boyer–Lindquist coordinates, where an unphysical, but mathematically more elegant radial coordinate r is used:

$$c^2 d\tau^2 = \left(1 - \frac{r_s r}{\rho^2}\right) c^2 dt^2 - \frac{\rho^2}{\Lambda^2} dr^2 - \rho^2 d\theta^2$$

$$-\left(r^2 + \alpha^2 + \frac{r_s r \alpha^2}{\rho^2} \sin^2\theta\right) \sin^2\theta \, d\phi^2 + \frac{2 r_s r \alpha c \sin^2\theta}{\rho^2} d\phi dt$$

where r_s is the Schwarzschild radius:

$$r_s = \frac{2GM}{c^2}$$

and where the following shorthand variables have been introduced for brevity:

$$\alpha = \frac{J}{Mc}$$

$$\rho^2 = r^2 + \alpha^2 \cos^2 \theta$$

$$\Lambda^2 = r^2 - r_s r + \alpha^2$$

In the non-relativistic limit where M (or, equivalently, rs) goes to zero, the Kerr metric becomes the orthogonal metric for the oblate spheroidal coordinates:

$$c^2 d\tau^2 = c^2 dt^2 - \frac{\rho^2}{r^2 + \alpha^2} dr^2 - \rho^2 d\theta^2 - \left(r^2 + \alpha^2\right)\sin^2 \theta d\phi^2$$

We may rewrite the Kerr metric in the following form:

$$c^2 d\tau^2 = \left(g_{tt} - \frac{g_{t\phi}^2}{g_{\phi\phi}}\right) dt^2 + g_{rr} dr^2 + g_{\theta\theta} d\theta^2 + g_{\phi\phi}\left(d\phi + \frac{g_{t\phi}}{g_{\phi\phi}} dt\right)^2$$

This metric is equivalent to a co-rotating reference frame that is rotating with angular speed Ω that depends on both the radius r and the colatitude θ:

$$\Omega = -\frac{g_{t\phi}}{g_{\phi\phi}} = \frac{r_s \alpha r c}{\rho^2 \left(r^2 + \alpha^2\right) + r_s \alpha^2 r \sin^2 \theta}$$

In the plane of the equator this simplifies to:

$$\Omega = \frac{r_s \alpha c}{r^3 + \alpha^2 r + r_s \alpha^2}$$

Thus, an inertial reference frame is entrained by the rotating central mass to participate in the latter's rotation; this is frame-dragging.

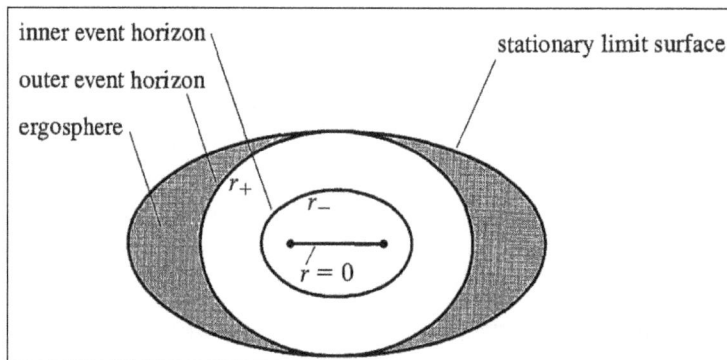

The two surfaces on which the Kerr metricappears to have singularities; the inner surface is the oblate spheroid-shaped event horizon, whereas the outer surface is pumpkin-shaped. The ergosphere lies between these two surfaces; within this volume, the purely temporal component g_{tt} is negative, i.e., acts like a purely spatial metric component. Consequently, particles within this ergosphere must co-rotate with the inner mass, if they are to retain their time-like character.

An extreme version of frame dragging occurs within the ergosphere of a rotating black hole. The Kerr metric has two surfaces on which it appears to be singular. The inner surface corresponds to a spherical event horizon similar to that observed in the Schwarzschild metric; this occurs at:

$$r_{inner} = \frac{r_s + \sqrt{r_s^2 - 4\alpha^2}}{2}$$

where the purely radial component grr of the metric goes to infinity. The outer surface can be approximated by an oblate spheroid with lower spin parameters, and resembles a pumpkin-shape with higher spin parameters. It touches the inner surface at the poles of the rotation axis, where the colatitude θ equals 0 or π; its radius in Boyer-Lindquist coordinates is defined by the formula:

$$r_{outer} = \frac{r_s + \sqrt{r_s^2 - 4\alpha^2 \cos^2 \theta}}{2}$$

where the purely temporal component gtt of the metric changes sign from positive to negative. The space between these two surfaces is called the ergosphere. A moving particle experiences a positive proper time along its worldline, its path through spacetime. However, this is impossible within the ergosphere, where gtt is negative, unless the particle is co-rotating with the interior mass M with an angular speed at least of Ω. However, as seen above, frame-dragging occurs about every rotating mass and at every radius r and colatitude θ, not only within the ergosphere.

Lense–Thirring Effect inside a Rotating Shell

The Lense–Thirring effect inside a rotating shell was taken by Albert Einstein as not just support for, but a vindication of Mach's principle, in a letter he wrote to Ernst Mach in 1913 (five years before Lense and Thirring's work, and two years before he had attained the final form of general relativity). A reproduction of the letter can be found in Misner, Thorne, Wheeler. The general effect scaled up to cosmological distances, is still used as a support for Mach's principle.

Inside a rotating spherical shell the acceleration due to the Lense–Thirring effect would be:

$$\bar{a} = -2d_1 \left(\bar{\omega} \times \bar{v} \right) - d_2 \left[\bar{\omega} \times \left(\bar{\omega} \times \bar{r} \right) + 2 \left(\overline{\omega r} \right) \bar{\omega} \right]$$

where the coefficients are:

$$d_1 = \frac{4MG}{3Rc^2}$$

$$d_2 = \frac{4MG}{15Rc^2}$$

for $MG \ll Rc^2$ or more precisely,

$$d_1 = \frac{4\alpha(2-\alpha)}{(1+\alpha)(3-\alpha)}, \qquad \alpha = \frac{MG}{2Rc^2}$$

The spacetime inside the rotating spherical shell will not be flat. A flat spacetime inside a rotating mass shell is possible if the shell is allowed to deviate from a precisely spherical shape and the mass density inside the shell is allowed to vary.

Lense–Thirring Precession

In general relativity, Lense–Thirring precession or the Lense–Thirring effect (named after Josef Lense and Hans Thirring) is a relativistic correction to the precession of a gyroscope near a large rotating mass such as the Earth. It is a gravitomagnetic frame-dragging effect. It is a prediction of general relativity consisting of secular precessions of the longitude of the ascending node and the argument of pericenter of a test particle freely orbiting a central spinning mass endowed with angular momentum S.

The difference between de Sitter precession and the Lense–Thirring effect is that the de Sitter effect is due simply to the presence of a central mass, whereas the Lense–Thirring effect is due to the rotation of the central mass. The total precession is calculated by combining the de Sitter precession with the Lense–Thirring precession.

According to a recent historical analysis by Pfister, the effect should be renamed as Einstein–Thirring–Lense effect.

The Lense–Thirring Metric

The gravitational field of a spinning spherical body of constant density was studied by Lense and Thirring in 1918, in the weak-field approximation. They obtained the metric:

$$ds^2 = \left(1 - \frac{2GM}{rc^2}\right)c^2\,dt^2 - \left(1 + \frac{2GM}{rc^2}\right)d\sigma^2 + 4G\epsilon_{ijk}S^k\frac{x^i}{c^3 r^3}c\,dt\,dx^j,$$

where the symbols are:

- ds^2 the metric,

- $d\sigma^2 = dx^2 + dy^2 + dz^2 = dr^2 + r^2 d\theta^2 + r^2\sin^2\theta\,d\varphi^2$ the flat-space line element in three dimensions,

- $r = \sqrt{x^2 + y^2 + z^2}$ the "radial" position of the observer,

- c the speed of light,

- G the gravitational constant,

- ϵ_{ijk} the completely antisymmetric Levi-Civita symbol,

- $M = \int T^{00}\,d^3x$ the mass of the rotating body,

- $S_k = \int \epsilon_{klm} x^l T^{m0} \, d^3x$ the angular momentum of the rotating body,

- $T^{\mu\nu}$ the energy-momentum tensor.

The above is the weak-field approximation of the full solution of the Einstein equations for a rotating body, known as the Kerr metric, which, due to the difficulty of its solution, was not obtained until 1965.

The Coriolis Term

The frame-dragging effect can be demonstrated in several ways. One way is to solve for geodesics; these will then exhibit a Coriolis force-like term, except that, in this case (unlike the standard Coriolis force), the force is not fictional, but is due to frame dragging induced by the rotating body. So, for example, an (instantaneously) radially infalling geodesic at the equator will satisfy the equation:

$$0 = r\frac{d^2\varphi}{dt^2} + 2\frac{GJ}{c^2 r^3}\frac{dr}{dt},$$

where:

- t is the time,

- φ is the azimuthal angle (longitudinal angle),

- $J = \| S \|$ is the magnitude of the angular momentum of the spinning massive body.

The above can be compared to the standard equation for motion subject to the Coriolis force:

$$0 = r\frac{d^2\varphi}{dt^2} + 2\omega\frac{dr}{dt},$$

where ω is the angular velocity of the rotating coordinate system. Note that, in either case, if the observer is not in radial motion, i.e. if $dr/dt = 0$, there is no effect on the observer.

Precession

The frame-dragging effect will cause a gyroscope to precess. The rate of precession is given by:

$$\Omega^k = \frac{G}{c^2 r^3}\left[S^k - 3\frac{(S\cdot x)x^k}{r^2}\right],$$

where:

- Ω is the angular velocity of the precession, a vector, and Ω_k one of its components,

- S_k the angular momentum of the spinning body, as before,

- $S\cdot x$ the ordinary flat-metric inner product of the position and the angular momentum.

That is, if the gyroscope's angular momentum relative to the fixed stars is L^i then it precesses as:

$$\frac{dL^i}{dt} = \epsilon_{ijk}\Omega^j L^k.$$

The rate of precession is given by:

$$\epsilon_{ijk}\Omega^k = \Gamma_{ij0},$$

where Γ_{ij0} is the Christoffel symbol for the above metric. "Gravitation" by Misner, Thorne, and Wheeler provides hints on how to most easily calculate this.

Gravitomagnetic Analysis

It is popular in some circles to use the gravitomagnetic approach to the linearized field equations. The reason for this popularity should be immediately evident below, by contrasting it to the difficulties of working with the equations above. The linearized metric $h_{\mu\nu} = g_{\mu\nu} - \eta_{\mu\nu}$ can be read off from the Lense–Thirring metric given above, where $ds^2 = g_{\mu\nu}\,dx^\mu\,dx^\nu ds^2 = g_{\mu\nu}\,dx^\mu\,dx^\nu$, and $\eta_{\mu\nu}\,dx^\mu\,dx^\nu = c^2\,dt^2 - dx^2 - dy^2 - dz^2$ In this approach, one writes the linearized metric, given in terms of the gravitomagneitc potentials ϕ and $\vec{A}$ is:

$$h_{00} = \frac{-2\phi}{c^2}$$

and

$$h_{0i} = \frac{2A_i}{c^2},$$

where:

$$\phi = \frac{-GM}{r}$$

is the gravito-electric potential, and:

$$\vec{A} = \frac{G}{r^3 c}\vec{S} \times \vec{r}$$

is the gravitomagnetic potential. Here $\vec{r}$ is the 3D spatial coordinate of the observer, and $\vec{S}$ is the angular momentum of the rotating body, exactly as defined above. The corresponding fields are:

$$\vec{E} = -\nabla\phi - \frac{1}{2c}\frac{\partial\vec{A}}{\partial t}$$

for the gravito-electric field, and:

$$\vec{B} = \frac{1}{2}\vec{\nabla} \times \vec{A}$$

is the gravitomagnetic field. It is then a matter of plugging and chugging to obtain:

$$\vec{B} = -\frac{G}{2cr^3}\left[\vec{S} - 3\frac{(\vec{S}\cdot\vec{r})\vec{r}}{r^2}\right]$$

as the gravitomagnetic field. Note that it is half the Lense–Thirring precession frequency. In this context, Lense–Thirring precession can essentially be viewed as a form of Larmor precession. The factor of 1/2 suggests that the correct gravitomagnetic analog of the gyromagnetic ratio is two.

The gravitomagnetic analog of the Lorentz force is given by:

$$\vec{F} = m\vec{E} + 4m\vec{v}\times\vec{B},$$

where m is the mass of a test particle moving with velocity $\vec{v}$. This can be used in a straightforward way to compute the classical motion of bodies in the gravitomagnetic field. For example, a radially infalling body will have a velocity $\vec{v} = -\hat{r}\,dr/dt$; direct substitution yields the Coriolis term.

Foucault's Pendulum

To get a sense of the magnitude of the effect, the above can be used to compute the rate of precession of Foucault's pendulum, located at the surface of the Earth.

For a solid ball of uniform density, such as the Earth, of radius R the moment of inertia is given by $2MR^2/5$, so that the absolute value of the angular momentum S is $\lvert S \rvert = 2MR^2\omega/5$, with ω the angular speed of the spinning ball.

The direction of the spin of the Earth may be taken as the z axis, whereas the axis of the pendulum is perpendicular to the Earth's surface, in the radial direction. Thus, we may take $\hat{z}\cdot\hat{r} = \cos\theta$ where θ is the latitude. Similarly, the location of the observer r is at the Earth's surface R. This leaves rate of precession is as:

$$\Omega_{LT} = \frac{2}{5}\frac{GM\omega}{c^2 R}\cos\theta.$$

As an example the latitude of the city of Nijmegen in the Netherlands is used for reference. This latitude gives a value for the Lense–Thirring precession:

$$\Omega_{LT} = 2.2{\cdot}10^{-4} \text{ arcseconds / day.}$$

At this rate a Foucault pendulum would have to oscillate for more than 16000 years to precess 1 degree. Despite being quite small, it is still two orders of magnitude larger than Thomas precession for such a pendulum.

The above does not include the de Sitter precession; it would need to be added to get the total relativistic precessions on Earth.

Experimental Verification

The Lense–Thirring effect, and the effect of frame dragging in general, continues to be studied experimentally. A rapidly changing orientation of a jet emitting from the black hole X-ray binary in V404 Cygni was observed. This jet has been modelled by "the Lense–Thirring precession of a vertically extended slim disk that arises from the super-Eddington accretion rate".

The Juno spacecraft's suite of science instruments will primarily characterize and explore the three-dimensional structure of Jupiter's polar magnetosphere, auroras and mass composition. As Juno is a polar-orbit mission, it will be possible to measure the orbital frame-dragging, known also as Lense–Thirring precession, caused by the angular momentum of Jupiter.

Astrophysical Setting

A star orbiting a spinning supermassive black hole experiences Lense–Thirring precession, causing its orbital line of nodes to precess at a rate:

$$\frac{d\Omega}{dt} = \frac{2GS}{c^2 a^3 \left(1-e^2\right)^{\frac{3}{2}}} = \frac{2G^2 M^2 \chi}{c^3 a^3 \left(1-e^2\right)^{\frac{3}{2}}},$$

where:

- a and e are the semimajor axis and eccentricity of the orbit,

- M is the mass of the black hole,

- χ is the dimensionless spin parameter ($0 < \chi < 1$).

Lense–Thirring precession of stars near the Milky Way supermassive black hole is expected to be measurable within the next few years.

The precessing stars also exert a torque back on the black hole, causing its spin axis to precess, at a rate:

$$\frac{dS}{dt} = \frac{2G}{c^2} \sum_j \frac{L_j \times S}{a_j^3 \left(1-e_j^2\right)^{\frac{3}{2}}},$$

where:

- L_j is the angular momentum of the j-th star,

- a_j and e_j are its semimajor axis and eccentricity.

A gaseous accretion disk that is tilted with respect to a spinning black hole will experience Lense–Thirring precession, at a rate given by the above equation, after setting $e = 0$ and identifying a with the disk radius. Because the precession rate varies with distance from the black hole, the disk will "wrap up", until viscosity forces the gas into a new plane, aligned with the black hole's spin axis (the "Bardeen–Petterson effect").

Geodetic Effect

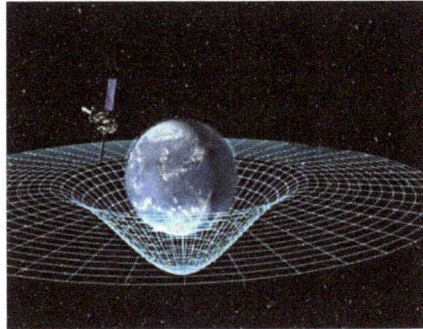

A representation of the geodetic effect.

The geodetic effect (also known as geodetic precession, de Sitter precession or de Sitter effect) represents the effect of the curvature of spacetime, predicted by general relativity, on a vector carried along with an orbiting body. For example, the vector could be the angular momentum of a gyroscope orbiting the Earth, as carried out by the Gravity Probe B experiment. The geodetic effect was first predicted by Willem de Sitter in 1916, who provided relativistic corrections to the Earth–Moon system's motion. De Sitter's work was extended in 1918 by Jan Schouten and in 1920 by Adriaan Fokker. It can also be applied to a particular secular precession of astronomical orbits, equivalent to the rotation of the Laplace–Runge–Lenz vector.

The term geodetic effect has two slightly different meanings as the moving body may be spinning or non-spinning. Non-spinning bodies move in geodesics, whereas spinning bodies move in slightly different orbits.

The difference between de Sitter precession and Lense–Thirring precession (frame dragging) is that the de Sitter effect is due simply to the presence of a central mass, whereas Lense–Thirring precession is due to the rotation of the central mass. The total precession is calculated by combining the de Sitter precession with the Lense–Thirring precession.

Experimental Confirmation

The geodetic effect was verified to a precision of better than 0.5% percent by Gravity Probe B, an experiment which measures the tilting of the spin axis of gyroscopes in orbit about the Earth. The first results were announced on April 14, 2007 at the meeting of the American Physical Society.

To derive the precession, assume the system is in a rotating Schwarzschild metric. The nonrotating metric is:

$$ds^2 = dt^2\left(1 - \frac{2m}{r}\right) - dr^2\left(1 - \frac{2m}{r}\right)^{-1} - r^2(d\theta^2 + \sin^2\theta\,d\phi'^2),$$

where $c = G = 1$.

We introduce a rotating coordinate system, with an angular velocity ω, such that a satellite in a circular orbit in the $\theta = \pi/2$ plane remains at rest. This gives us:

$$d\phi = d\phi' - \omega \, dt.$$

In this coordinate system, an observer at radial position r sees a vector positioned at r as rotating with angular frequency ω. This observer, however, sees a vector positioned at some other value of r as rotating at a different rate, due to relativistic time dilation. Transforming the Schwarzschild metric into the rotating frame, and assuming that θ is a constant, we find:

$$ds^2 = \left(1 - \frac{2m}{r} - r^2 \beta \omega^2\right)\left(dt - \frac{r^2 \beta \omega}{1 - 2m/r - r^2 \beta \omega^2} d\phi\right)^2 -$$
$$- dr^2 \left(1 - \frac{2m}{r}\right)^{-1} - \frac{r^2 \beta - 2mr\beta}{1 - 2m/r - r^2 \beta \omega^2} d\phi^2,$$

with $\beta = \sin^2(\theta)$. For a body orbiting in the $\theta = \pi/2$ plane, we will have $\beta = 1$, and the body's world-line will maintain constant spatial coordinates for all time. Now, the metric is in the canonical form:

$$ds^2 = e^{2\Phi}\left(dt - w_i \, dx^i\right)^2 - k_{ij} \, dx^i \, dx^j.$$

From this canonical form, we can easily determine the rotational rate of a gyroscope in proper time:

$$\Omega = \frac{\sqrt{2}}{4} e^{\Phi} [k^{ik} k^{jl} (\omega_{i,j} - \omega_{j,i})(\omega_{k,l} - \omega_{l,k})]^{1/2} =$$
$$= \frac{\sqrt{\beta}\omega(r - 3m)}{r - 2m - \beta \omega^2 r^3} = \sqrt{\beta}\omega.$$

where the last equality is true only for free falling observers for which there is no acceleration, and thus $\Phi_{,i} = 0$. This leads to:

$$\Phi_{,i} = \frac{2m/r^2 - 2r\beta\omega^2}{2(1 - 2m/r - r^2 \beta \omega^2)} = 0.$$

Solving this equation for ω yields:

$$\omega^2 = \frac{m}{r^3 \beta}.$$

This is essentially Kepler's law of periods, which happens to be relativistically exact when expressed in terms of the time coordinate t of this particular rotating coordinate system. In the rotating frame, the satellite remains at rest, but an observer aboard the satellite sees the gyroscope's angular momentum vector precessing at the rate ω. This observer also sees the distant stars as rotating, but they rotate at a slightly different rate due to time dilation. Let τ be the gyroscope's proper time. Then:

$$\Delta\tau = \left(1 - \frac{2m}{r} - r^2\beta\omega^2\right)^{1/2} dt = \left(1 - \frac{3m}{r}\right)^{1/2} dt.$$

The $-2m/r$ term is interpreted as the gravitational time dilation, while the additional $-m/r$ is due to the rotation of this frame of reference. Let α' be the accumulated precession in the rotating frame. Since $\alpha' = \Omega\Delta\tau$, the precession over the course of one orbit, relative to the distant stars, is given by:

$$\alpha = \alpha' + 2\pi = -2\pi\sqrt{\beta}\left(\left(1 - \frac{3m}{r}\right)^{1/2} - 1\right).$$

with a first-order Taylor series we find:

$$\alpha \approx \frac{3\pi m}{r}\sqrt{\beta} = \frac{3\pi m}{r}\sin(\theta).$$

Gravitational Singularity

A gravitational singularity (or space-time singularity) is a location where the quantities that are used to measure the gravitational field become infinite in a way that does not depend on the co-ordinate system. In other words, it is a point in which all physical laws are indistinguishable from one another, where space and time are no longer interrelated realities, but merge indistinguishably and cease to have any independent meaning.

This artist's impression depicts a rapidly spinning supermassive black hole surrounded by an accretion disc.

Singularities were first predicated as a result of Einstein's Theory of General Relativity, which resulted in the theoretical existence of black holes. In essence, the theory predicted that any star reaching beyond a certain point in its mass (aka. the Schwarzschild Radius) would exert a gravitational force so intense that it would collapse.

At this point, nothing would be capable of escaping its surface, including light. This is due to the fact the gravitational force would exceed the speed of light in vacuum – 299,792,458 meters per second (1,079,252,848.8 km/h; 670,616,629 mph).

This phenomena is known as the Chandrasekhar Limit, named after the Indian astrophysicist Subrahmanyan Chandrasekhar, who proposed it in 1930. At present, the accepted value of this limit is believed to be 1.39 Solar Masses (i.e. 1.39 times the mass of our Sun), which works out to a whopping 2.765×10^{30} kg (or 2,765 trillion trillion metric tons).

Another aspect of modern General Relativity is that at the time of the Big Bang (i.e. the initial state of the Universe) was a singularity. Roger Penrose and Stephen Hawking both developed theories that attempted to answer how gravitation could produce singularities, which eventually merged together to be known as the Penrose–Hawking Singularity Theorems.

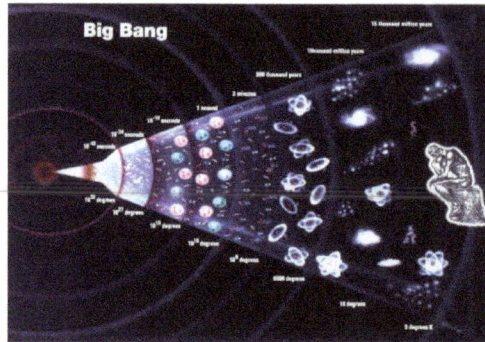

The Big Bang Theory: A history of the Universe starting from a singularity and expanding ever since.

According to the Penrose Singularity Theorem, which he proposed in 1965, a time-like singularity will occur within a black hole whenever matter reaches certain energy conditions. At this point, the curvature of space-time within the black hole becomes infinite, thus turning it into a trapped surface where time ceases to function.

The Hawking Singularity Theorem added to this by stating that a space-like singularity can occur when matter is forcibly compressed to a point, causing the rules that govern matter to break down. Hawking traced this back in time to the Big Bang, which he claimed was a point of infinite density. However, Hawking later revised this to claim that general relativity breaks down at times prior to the Big Bang, and hence no singularity could be predicted by it.

Some more recent proposals also suggest that the Universe did not begin as a singularity. These includes theories like Loop Quantum Gravity, which attempts to unify the laws of quantum physics with gravity. This theory states that, due to quantum gravity effects, there is a minimum distance beyond which gravity no longer continues to increase, or that interpenetrating particle waves mask gravitational effects that would be felt at a distance.

Types of Singularities

The two most important types of space-time singularities are known as Curvature Singularities and Conical Singularities. Singularities can also be divided according to whether they are covered by an event horizon or not. In the case of the former, you have the Curvature and Conical; whereas in the latter, you have what are known as Naked Singularities.

A Curvature Singularity is best exemplified by a black hole. At the center of a black hole, space-time becomes a one-dimensional point which contains a huge mass. As a result, gravity become infinite and space-time curves infinitely, and the laws of physics as we know them cease to function.

Conical singularities occur when there is a point where the limit of every general covariance quantity is finite. In this case, space-time looks like a cone around this point, where the singularity is located at the tip of the cone. An example of such a conical singularity is a cosmic string, a type of hypothetical one-dimensional point that is believed to have formed during the early Universe.

And, as mentioned, there is the Naked Singularity, a type of singularity which is not hidden behind an event horizon. These were first discovered in 1991 by Shapiro and Teukolsky using computer simulations of a rotating plane of dust that indicated that General Relativity might allow for "naked" singularities.

In this case, what actually transpires within a black hole (i.e. its singularity) would be visible. Such a singularity would theoretically be what existed prior to the Big Bang. The key word here is theoretical, as it remains a mystery what these objects would look like.

For the moment, singularities and what actually lies beneath the veil of a black hole remains a mystery. As time goes on, it is hoped that astronomers will be able to study black holes in greater detail. It is also hoped that in the coming decades, scientists will find a way to merge the principles of quantum mechanics with gravity, and that this will shed further light on how this mysterious force operates.

Gravitational Time Dilation

Gravitational time dilation is a form of time dilation, an actual difference of elapsed time between two events as measured by observers situated at varying distances from a gravitating mass. The higher the gravitational potential (the farther the clock is from the source of gravitation), the faster time passes. Albert Einstein originally predicted this effect in his theory of relativity and it has since been confirmed by tests of general relativity.

This has been demonstrated by noting that atomic clocks at differing altitudes (and thus different gravitational potential) will eventually show different times. The effects detected in such Earth-bound experiments are extremely small, with differences being measured in nanoseconds. Relative to Earth's age in billions of years, Earth's core is effectively 2.5 years younger than its surface. Demonstrating larger effects would require greater distances from the Earth or a larger gravitational source.

Gravitational time dilation was first described by Albert Einstein in 1907 as a consequence of special relativity in accelerated frames of reference. In general relativity, it is considered to be a difference in the passage of proper time at different positions as described by a metric tensor of space-time. The existence of gravitational time dilation was first confirmed directly by the Pound–Rebka experiment in 1959.

Clocks that are far from massive bodies (or at higher gravitational potentials) run more quickly, and clocks close to massive bodies (or at lower gravitational potentials) run more slowly. For example, considered over the total time-span of Earth (4.6 billion years), a clock set at the peak of Mount Everest would be about 39 hours ahead of a clock set at sea level. This is because gravitational time dilation is manifested in accelerated frames of reference or, by virtue of the equivalence principle, in the gravitational field of massive objects.

According to general relativity, inertial mass and gravitational mass are the same, and all accelerated reference frames (such as a uniformly rotating reference frame with its proper time dilation) are physically equivalent to a gravitational field of the same strength.

Consider a family of observers along a straight "vertical" line, each of whom experiences a distinct constant g-force directed along this line (e.g., a long accelerating spacecraft, a skyscraper, a shaft on a planet). Let g(h) be the dependence of g-force on "height", a coordinate along the aforementioned line. The equation with respect to a base observer at h = 0 is:

$$T_d(h) = \exp\left[\frac{1}{c^2}\int_0^h g(h')dh'\right]$$

where $T_d(h)$ is the total time dilation at a distant position h, $g(h)$ is the dependence of g-force on "height" c is the speed of light, and exp denotes exponentiation by e.

For simplicity, in a Rindler's family of observers in a flat space-time, the dependence would be:

$$g(h) = c^2 / (H+h)$$

with constant H, which yields:

$$T_d(h) = e^{\ln(H+h)-\ln H} = \frac{H+h}{h}.$$

On the other hand, when g is nearly constant and gh is much smaller than c², the linear "weak field" approximation $T_d = 1 + gh/c^2$ can also be used.

Outside a Non-rotating Sphere

A common equation used to determine gravitational time dilation is derived from the Schwarzschild metric, which describes space-time in the vicinity of a non-rotating massive spherically symmetric object. The equation is:

$$t_0 = t_f\sqrt{1-\frac{2GM}{rc^2}} = t_f\sqrt{1-\frac{r_s}{r}}$$

where:

- t_0 is the proper time between events A and B for a slow-ticking observer within the gravitational field,

- t_f is the coordinate time between events A and B for a fast-ticking observer at an arbitrarily large distance from the massive object (this assumes the fast-ticking observer is using Schwarzschild coordinates, a coordinate system where a clock at infinite distance from the massive sphere would tick at one second per second of coordinate time, while closer clocks would tick at less than that rate),

- G is the gravitational constant,

- M is the mass of the object creating the gravitational field,

- r is the radial coordinate of the observer (which is analogous to the classical distance from the center of the object, but is actually a Schwarzschild coordinate),

- c is the speed of light,

- r $2GM / c$ is the Schwarzschild radius of M.

To illustrate then, without accounting for the effects of rotation, proximity to Earth's gravitational well will cause a clock on the planet's surface to accumulate around 0.0219 fewer seconds over a period of one year than would a distant observer's clock. In comparison, a clock on the surface of the sun will accumulate around 66.4 fewer seconds in one year.

Circular Orbits

In the Schwarzschild metric, free-falling objects can be in circular orbits if the orbital radius is larger than $\frac{3}{2} r_s$ (the radius of the photon sphere). The formula for a clock at rest is given above; the formula below gives the gravitational time dilation for a clock in a circular orbit but it does not include the opposing time dilation caused by the clock's motion:

$$t_0 = t_f \sqrt{1 - \frac{3}{2} \frac{r_s}{r}}.$$

Important Features of Gravitational Time Dilation

- According to the general theory of relativity, gravitational time dilation is copresent with the existence of an accelerated reference frame. Additionally, all physical phenomena in similar circumstances undergo time dilation equally according to the equivalence principle used in the general theory of relativity.

- The speed of light in a locale is always equal to c according to the observer who is there. That is, every infinitesimal region of space time may be assigned its own proper time and the speed of light according to the proper time at that region is always c. This is the case whether or not a given region is occupied by an observer. A time delay can be measured for photons which are emitted from Earth, bend near the Sun, travel to Venus, and then return to Earth along a similar path. There is no violation of the constancy of the speed of light here, as any observer observing the speed of photons in their region will find the speed of those photons to be c, while the speed at which we observe light travel finite distances in the vicinity of the Sun will differ from c.

- If an observer is able to track the light in a remote, distant locale which intercepts a remote, time dilated observer nearer to a more massive body, that first observer tracks that both the remote light and that remote time dilated observer have a slower time clock than other light which is coming to the first observer at c, like all other light the first observer *really* can observe (at their own location). If the other, remote light eventually intercepts the first observer, it too will be measured at c by the first observer.

- Time dilation in a gravitational field is equal to time dilation in far space, due to a speed that is needed to escape that gravitational field. Here is the proof:

  ○ Time dilation inside a gravitational field g is $t_0 = t_f \sqrt{1 - \dfrac{2GM}{rc^2}}$.

  ○ Escape velocity from g is $\sqrt{2GM / r}$.

  ○ Time dilation formula per special relativity is $t = t \sqrt{1 - v^2 / c^2}$.

  ○ Substituting escape velocity for v in the above $t_0 = t_f \sqrt{1 - \dfrac{2GM}{rc^2}}$.

This should be true for any gravitational fields considering simple scenarios like non-rotation etc.

Below are some evident examples when substituting $2GM / c^2$ by the Schwarzschild radius, giving $t_0 = t_f \sqrt{1 - r_s / r}$:

- Time stops at surface of a black hole since $r = r_s$ gives $t_0 = t_f \sqrt{1 - r_s / r} = t_f \sqrt{1 - r / r} = 0$.

- Escape velocity from surface of a black hole is c, since replacing Schwarzschild radius gives $v = \sqrt{2GMc^2 / 2GM} = \sqrt{c^2} = c$.

- Time stops at speed c, since replacing v gives $t_0 = t_f \sqrt{1 - c^2 / c^2} = 0$.

where:

- t_0 is the proper time between events A and B for a slow-ticking observer within the gravitational field.

- t_f is the coordinate time between events A and B for a fast-ticking observer at an arbitrarily large distance from the massive object.

- G is the Gravitational constant.

- M is the mass of the object creating the gravitational field.

- r is the radial coordinate of the observer which is analogous to the classical distance from the center of the object.

- c is the speed of light.

- v is the velocity.

- g is gravitational acceleration/field = GM / r^2.

Experimental Confirmation

Gravitational time dilation has been experimentally measured using atomic clocks on airplanes. The clocks aboard the airplanes were slightly faster than clocks on the ground. The effect is significant enough that the Global Positioning System's artificial satellites need to have their clocks corrected.

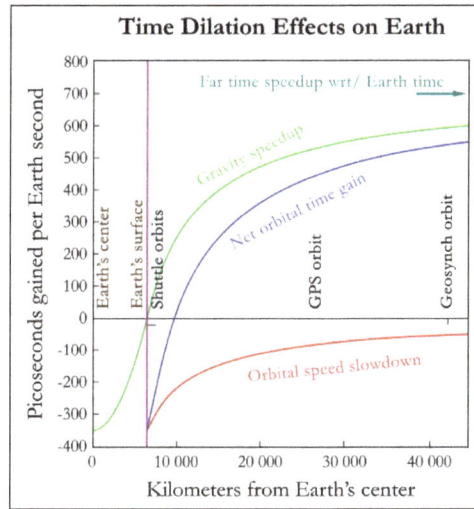

Satellite clocks are slowed by their orbital speed, but accelerated
by their distance out of Earth's gravitational well.

Additionally, time dilations due to height differences of less than one metre have been experimentally verified in the laboratory.

Gravitational time dilation has also been confirmed by the Pound–Rebka experiment, observations of the spectra of the white dwarf Sirius B, and experiments with time signals sent to and from Viking 1 Mars lander.

Gravitational Redshift

In Einstein's general theory of relativity, the gravitational redshift is the phenomenon that clocks in a gravitational field tick slower when observed by a distant observer. More specifically the term refers to the shift of wavelength of a photon to longer wavelength (the red side in an optical spectrum) when observed from a point in a lower gravitational field. In the latter case the 'clock' is the frequency of the photon and a lower frequency is the same as a longer ("redder") wavelength.

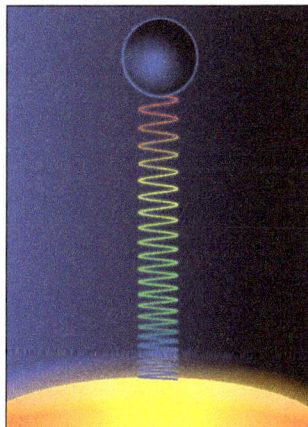

The gravitational redshift of a light wave as it moves upwards against a gravitational field (produced by the yellow star below). The effect is greatly exaggerated in this diagram.

The gravitational redshift is a simple consequence of Einstein's equivalence principle ("all bodies fall with the same acceleration, independent of their composition") and was found by Einstein eight years before the full theory of relativity.

Observing the gravitational redshift in the solar system is one of the classical tests of general relativity. Gravitational redshifts are an important effect in satellite-based navigation systems such as GPS. If the effects of general relativity were not taken into account, such systems would not work at all.

Prediction by the Equivalence Principle and General Relativity

Einstein's theory of general relativity incorporates the equivalence principle, which can be stated in various different ways. One such statement is that gravitational effects are locally undetectable for a free-falling observer. Therefore, in a laboratory experiment at the surface of the earth, all gravitational effects should be equivalent to the effects that would have been observed if the laboratory had been accelerating through outer space at g. One consequence is a gravitational Doppler effect. If a light pulse is emitted at the floor of the laboratory, then a free-falling observer says that by the time it reaches the ceiling, the ceiling has accelerated away from it, and therefore when observed by a detector fixed to the ceiling, it will be observed to have been Doppler shifted toward the red end of the spectrum. This shift, which the free-falling observer considers to be a kinematical Doppler shift, is thought of by the laboratory observer as a gravitational redshift. Such an effect was verified in the 1959 Pound–Rebka experiment. In a case such as this, where the gravitational field is uniform, the change in wavelength is given by:

$$\frac{\Delta\lambda}{\lambda} \approx \frac{g\Delta y}{c^2},$$

Where Δy is the change in height. Since this prediction arises directly from the equivalence principle, it does not require any of the mathematical apparatus of general relativity, and its verification does not specifically support general relativity over any other theory that incorporates the equivalence principle.

When the field is not uniform, the simplest and most useful case to consider is that of a spherically symmetric field. By Birkhoff's theorem, such a field is described in general relativity by the Schwarzschild metric,

$d\tau^2 = (1 - r_s/R)dt^2 + \ldots$ where $d\tau$ is the clock time of an observer at distance R from the center, dt

is the time measured by an observer at infinity, r_s is the Schwarzschild radius $2GM/c^2$, represents terms that vanish if the observer is at rest, G is Newton's gravitational constant, M the mass of the gravitating body, and c the speed of light. The result is that frequencies and wavelengths are shifted according to the ratio:

$$\frac{\lambda_\infty}{\lambda_e} = (1 - r_s/R_e)^{-1/2},$$

Where λ_∞ is the wavelength of the light as measured by the observer at infinity, λ_e is the wavelength measured at the source of emission, and R_e radius at which the photon is emitted. This can be related to the redshift parameter conventionally defined as $z = \lambda_\infty / \lambda_e - 1$ In the case where neither the emitter nor the observer is at infinity, the transitivity of Doppler shifts allows us to generalize the result to $\lambda_1 / \lambda_2 = [(1 - r_s / R_1)/(1 - r_s / R_2)]^{1/2}$ The redshift formula for the frequency $\nu = c / \lambda$ is $\nu_o / \nu_e = \lambda_e / \lambda_o$. When $R_1 - R_2, R_1 - R_2$ is small, these results are consistent with the equation given above based on the equivalence principle.

For an object compact enough to have an event horizon, the redshift is not defined for photons emitted inside the Schwarzschild radius, both because signals cannot escape from inside the horizon and because an object such as the emitter cannot be stationary inside the horizon, as was assumed above. Therefore, this formula only applies when R_e is larger than r_s When the photon is emitted at a distance equal to the Schwarzschild radius, the redshift will be infinitely large, and it will not escape to any finite distance from the Schwarzschild sphere. When the photon is emitted at an infinitely large distance, there is no redshift.

In the Newtonian limit, i.e. when R_e is sufficiently large compared to the Schwarzschild radius r_s the redshift can be approximated as:

$$z \approx \frac{1}{2} \frac{r_s}{R_e} = \frac{GM}{c^2 R_e}$$

Experimental Verification

Initial Observations of Gravitational Redshift of White Dwarf Stars

A number of experimenters initially claimed to have identified the effect using astronomical measurements, and the effect was considered to have been finally identified in the spectral lines of the star Sirius B by W.S. Adams in 1925. However, measurements by Adams have been criticized as being too low and these observations are now considered to be measurements of spectra that are unusable because of scattered light from the primary, Sirius A. The first accurate measurement of the gravitational redshift of a white dwarf was done by Popper in 1954, measuring a 21 km/sec gravitational redshift of 40 Eridani B.

The redshift of Sirius B was finally measured by Greenstein et al. in 1971, obtaining the value for the gravitational redshift of 89±19 km/sec, with more accurate measurements by the Hubble Space Telescope, showing 80.4±4.8 km/sec.

Terrestrial Tests

The effect is now considered to have been definitively verified by the experiments of Pound, Rebka and Snider between 1959 and 1965. The Pound–Rebka experiment of 1959 measured the gravitational redshift in spectral lines using a terrestrial ^{57}Fe gamma source over a vertical height of 22.5 metres. using measurements of the change in wavelength of gamma-ray photons generated with the Mössbauer effect, which generates radiation with a very narrow line width. The accuracy of the gamma-ray measurements was typically 1%.

An improved experiment was done by Pound and Snider in 1965, with an accuracy better than the 1% level.

A very accurate gravitational redshift experiment was performed in 1976, where a hydrogen maser clock on a rocket was launched to a height of 10,000 km, and its rate compared with an identical clock on the ground. It tested the gravitational redshift to 0.007%.

Later tests can be done with the Global Positioning System (GPS), which must account for the gravitational redshift in its timing system, and physicists have analyzed timing data from the GPS to confirm other tests. When the first satellite was launched, it showed the predicted shift of 38 microseconds per day. This rate of the discrepancy is sufficient to substantially impair the function of GPS within hours if not accounted for. An excellent account of the role played by general relativity in the design of GPS can be found in Ashby 2003.

Later Astronomical Measurements

James W. Brault, a graduate student of Robert Dicke at Princeton University, measured the gravitational redshift of the sun using optical methods in 1962.

In 2011 the group of Radek Wojtak of the Niels Bohr Institute at the University of Copenhagen collected data from 8000 galaxy clusters and found that the light coming from the cluster centers tended to be red-shifted compared to the cluster edges, confirming the energy loss due to gravity.

Other precision tests of general relativity, not discussed here, are the Gravity Probe A satellite, launched in 1976, which showed gravity and velocity affect the ability to synchronize the rates of clocks orbiting a central mass; the Hafele–Keating experiment, which used atomic clocks in circumnavigating aircraft to test general relativity and special relativity together; and the forthcoming Satellite Test of the Equivalence Principle.

In 2018, the VLT had successfully observed the gravitational redshift and the first successful test has been performed by the Galactic Centre team at the Max Planck Institute for Extraterrestrial Physics (MPE).

Development of the Theory

The gravitational weakening of light from high-gravity stars was predicted by John Michell in 1783 and Pierre-Simon Laplace in 1796, using Isaac Newton's concept of light corpuscles and who predicted that some stars would have a gravity so strong that light would not be able to escape. The effect of gravity on light was then explored by Johann Georg von Soldner, who calculated the amount of deflection of a light ray by the sun, arriving at the Newtonian answer which is half the value predicted by general relativity. All of this early work assumed that light could slow down and fall, which was inconsistent with the modern understanding of light waves.

Once it became accepted that light was an electromagnetic wave, it was clear that the frequency of light should not change from place to place, since waves from a source with a fixed frequency keep the same frequency everywhere. One way around this conclusion would be if time itself were altered—if clocks at different points had different rates.

This was precisely Einstein's conclusion in 1911. He considered an accelerating box, and noted that according to the special theory of relativity, the clock rate at the "bottom" of the box (the side away from the direction of acceleration) was slower than the clock rate at the "top" (the side toward the direction of acceleration). Nowadays, this can be easily shown in accelerated coordinates. The metric tensor in units where the speed of light is one is:

$$ds^2 = -r^2 dt^2 + dr^2$$

and for an observer at a constant value of r, the rate at which a clock ticks, R(r), is the square root of the time coefficient, R(r)=r. The acceleration at position r is equal to the curvature of the hyperbola at fixed r, and like the curvature of the nested circles in polar coordinates, it is equal to 1/r.

So at a fixed value of g, the fractional rate of change of the clock-rate, the percentage change in the ticking at the top of an accelerating box vs at the bottom, is:

$$\frac{R(r+dr)-R(r)}{R} = \frac{dr}{r} = gdr$$

The rate is faster at larger values of R, away from the apparent direction of acceleration. The rate is zero at r=0, which is the location of the acceleration horizon.

Using the equivalence principle, Einstein concluded that the same thing holds in any gravitational field, that the rate of clocks R at different heights was altered according to the gravitational field g. When g is slowly varying, it gives the fractional rate of change of the ticking rate. If the ticking rate is everywhere almost this same, the fractional rate of change is the same as the absolute rate of change, so that:

$$\frac{dR}{dx} = g = -\frac{dV}{dx}$$

Since the rate of clocks and the gravitational potential have the same derivative, they are the same up to a constant. The constant is chosen to make the clock rate at infinity equal to 1. Since the gravitational potential is zero at infinity:

$$R(x) = 1 - \frac{V(x)}{c^2}$$

where the speed of light has been restored to make the gravitational potential dimensionless.

The coefficient of the dt^2 in the metric tensor is the square of the clock rate, which for small values of the potential is given by keeping only the linear term:

$$R^2 = 1 - 2V$$

and the full metric tensor is:

$$ds^2 = -\left(1 - \frac{2V(r)}{c^2}\right)c^2 dt^2 + dx^2 + dy^2 + dz^2$$

where again the C's have been restored. This expression is correct in the full theory of general relativity, to lowest order in the gravitational field, and ignoring the variation of the space-space and space-time components of the metric tensor, which only affect fast moving objects.

Using this approximation, Einstein reproduced the incorrect Newtonian value for the deflection of light in 1909. But since a light beam is a fast moving object, the space-space components contribute too. After constructing the full theory of general relativity in 1916, Einstein solved for the space-space components in a post-Newtonian approximation and calculated the correct amount of light deflection – double the Newtonian value. Einstein's prediction was confirmed by many experiments, starting with Arthur Eddington's 1919 solar eclipse expedition.

The changing rates of clocks allowed Einstein to conclude that light waves change frequency as they move, and the frequency/energy relationship for photons allowed him to see that this was best interpreted as the effect of the gravitational field on the mass–energy of the photon. To calculate the changes in frequency in a nearly static gravitational field, only the time component of the metric tensor is important, and the lowest order approximation is accurate enough for ordinary stars and planets, which are much bigger than their Schwarzschild radius.

Shapiro Time Delay

The Shapiro time delay effect, or gravitational time delay effect, is one of the four classic solar-system tests of general relativity. Radar signals passing near a massive object take slightly longer to travel to a target and longer to return than they would if the mass of the object were not present. The time delay is caused by spacetime dilation, which increases the path length. In an article entitled Fourth Test of General Relativity, astrophysicist Irwin Shapiro wrote:

> "Because, according to the general theory, the speed of a light wave depends on the strength of the gravitational potential along its path, these time delays should thereby be increased by almost $2 \times 10-4$ sec when the radar pulses pass near the sun. Such a change, equivalent to 60 km in distance, could now be measured over the required path length to within about 5 to 10% with presently obtainable equipment."

Shapiro uses c as the speed of light and calculated the time delay of the passage of light waves or rays over finite coordinate distance according to a Schwarzschild solution to the Einstein field equations.

The time delay effect was first predicted in 1964, by Irwin Shapiro. Shapiro proposed an observational test of his prediction: bounce radar beams off the surface of Venus and Mercury and measure the round-trip travel time. When the Earth, Sun, and Venus are most favorably aligned, Shapiro showed that the expected time delay, due to the presence of the Sun, of a radar signal traveling from the Earth to Venus and back, would be about 200 microseconds, well within the limitations of 1960s-era technology.

The first tests, performed in 1966 and 1967 using the MIT Haystack radar antenna, were successful, matching the predicted amount of time delay. The experiments have been repeated many times since then, with increasing accuracy.

Calculating Time Delay

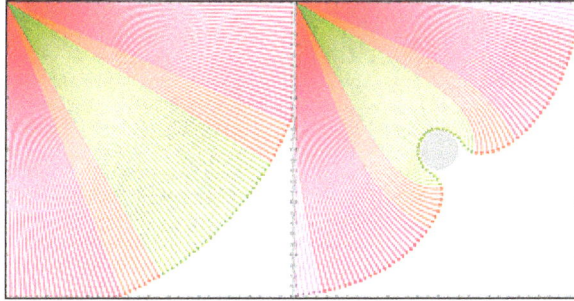

Left: unperturbed lightrays in a flat spacetime, right: shapiro-delayed and
deflected lightrays in the vicinity of a gravitating mass.

In a nearly static gravitational field of moderate strength (say, of stars and planets, but not one of
a black hole or close binary system of neutron stars) the effect may be considered as a special case
of gravitational time dilation. The measured elapsed time of a light signal in a gravitational field is
longer than it would be without the field, and for moderate-strength nearly static fields the difference is directly proportional to the classical gravitational potential, precisely as given by standard
gravitational time dilation formulas.

Time Delay due to Light Traveling around a Single Mass

Shapiro's original formulation was derived from the Schwarzschild solution and included terms
to the first order in solar mass (M) for a proposed Earth-based radar pulse bouncing off an inner
planet and returning passing close to the Sun:

$$\Delta t \approx \frac{4GM}{c^3}\left\{\ln\left[\frac{x_p + (x_p^2 + d^2)^{1/2}}{-x_e + (x_e^2 + d^2)^{1/2}}\right] - \frac{1}{2}\left[\frac{x_p}{(x_p^2 + d^2)^{1/2}} + \frac{x_e}{(x_e^2 + d^2)^{1/2}}\right]\right\} + O\left(\frac{G^2M^2}{c^6}\right),$$

where d is the distance of closest approach of the radar wave to the center of the Sun, x_e is the distance along the line of flight from the Earth-based antenna to the point of closest approach to the
Sun, and x_p represents the distance along the path from this point to the planet. The right-hand
side of this equation is primarily due to the variable speed of the light ray; the contribution from
the change in path, being of second order in M, is negligible. In the limit when the distance of
closest approach is much larger than the Schwarzschild radius, relativistic Newtonian dynamics
predicts:

$$\Delta t \approx \frac{2GM}{c^3}\ln\frac{4x_p x_e}{d^2},$$

which agrees with the known formula for the Shapiro time delay quoted in the literature derived
using general relativity.

For a signal going around a massive object, the time delay can be calculated as the following:

$$\Delta t = -\frac{2GM}{c^3}\ln(1 - R\cdot x).$$

Here R is the unit vector pointing from the observer to the source, and x is the unit vector pointing from the observer to the gravitating mass M. The dot denotes the usual Euclidean dot product.

Using $\Delta x = c\Delta t$, this formula can also be written as:

$$\Delta x = -R_s \ln(1 - R \cdot x),$$

which is the extra distance the light has to travel. Here R_s is the Schwarzschild radius.

In PPN parameters,

$$\Delta t = -(1 + \gamma)\frac{R_s}{2c}\ln(1 - R \cdot x),$$

which is twice the Newtonian prediction (with $\gamma = 0$).

Interplanetary Probes

Shapiro delay must be considered along with ranging data when trying to accurately determine the distance to interplanetary probes such as the Voyager and Pioneer spacecraft.

Shapiro Delay of Neutrinos and Gravitational Waves

From the nearly simultaneous observations of neutrinos and photons from SN 1987A, the Shapiro delay for high-energy neutrinos must be the same as that for photons to within 10%, consistent with recent estimates of the neutrino mass, which imply that those neutrinos were moving at very close to the speed of light. After the direct detection of gravitational waves in 2016, the one-way Shapiro delay was calculated by two groups and is about 1800 days. In general relativity and other metric theories of gravity, though, the Shapiro delay for gravitational waves is expected to be the same as that for light and neutrinos. However, in theories such as tensor-vector-scalar gravity and other modified GR theories, which reproduce Milgrom's law and avoid the need for dark matter, the Shapiro delay for gravitational waves is much smaller than that for neutrinos or photons. The observed 1.7-second difference in arrival times seen between gravitational wave and gamma ray arrivals from neutron star merger GW170817 was far less than the estimated Shapiro delay of about 1000 days. This rules out a class of modified models of gravity that dispense with the need for dark matter.

Gravitational Wave

Gravitational waves are distortions in the fabric of space and time caused by the movement of massive objects, like sound waves in air or the ripples made on a pond's surface when someone throws a rock in the water. But unlike sound waves pond ripples, which spread out through a medium like watter, gravitational waves are vibrations in spacetime itself, which means they move just fine through the vacuum of space. And unlike the gentle drop of a stone in a pond, the events that trigger gravitational waves are among the most powerful in the universe.

We can hear gravitational waves, in the same sense that sound waves travel through water, or seismic waves move through the earth. The difference is that sound waves vibrate through a medium, like water or soil. For gravitational waves, spacetime is the medium. It just takes the right instrument to hear them.

Detecting gravitational waves on Earth was a challenge that took roughly a century to complete, since the ones that wash through the planet are incredibly tiny.

Detecting Gravitational Waves

Einstein's general theory of relativity first predicted the existence of gravitational waves, which the famous scientist himself noted in 1916. Though Einstein later doubted the waves' existence, we have had indirect evidence of them since the 1970s.

In 1974, astronomers Joe Taylor and Russell Hulse tracked a pair of spinning stellar corpses called pulsars. As the pair of pulsars spun around each other, they grew closer together, which indicated that they were giving off energy. Calculations made clear that this energy loss came in the form of gravitational waves—a discovery that won Taylor and Hulse a Nobel Prize in 1993.

The first direct detection of gravitational waves took place on September 14, 2015, when the U.S. Laser Interferometry Gravitational-Wave Observatory—aka LIGO—detected the rumble that two colliding black holes gave off 1.3 billion years ago. Scientists formally announced the success in February 2016. In 2017, three of LIGO's founding scientists were honored with the Nobel Prize in physics.

Starting in the 1970s, physicists including Rainer Weiss, Kip Thorne, and Barry Barish sketched out the idea that later became LIGO. The observatory consists of two facilities: one in Louisiana, the other in Washington State. Each L-shaped facility consists of two arms more than two miles long that meet at a right angle.

By bouncing lasers back and forth within each arm, physicists can measure their lengths with an accuracy so astonishing, it would be like measuring the distance between us and Alpha Centauri—the closest star outside our solar system—to within a hair's width. When a gravitational wave passes through Earth, it slightly stretches one of the arms and compresses the other. Those length changes alter the time it takes the laser beams to bounce back and forth, which in turn changes the pattern the beams make where they meet. By tracking the shifting patterns through time, researchers can watch a gravitational wave ripple through the facility.

LIGO has two facilities so that both detectors can try and spot the same event, in effect checking each other's work. In addition, the difference in time between each detection reveals which direction the gravitational waves came from, helping astronomers hoping to pinpoint the source in the sky.

Information from Gravitational Waves

The analogy that some physicists use is that gravitational waves let us "hear the universe." To be clear, sound and gravitational waves are very different things. But by watching events play out in the universe at different wavelengths of light, while also watching out for the vibrations of gravitational waves, we can embark on what's known as multi-messenger astronomy.

Today's gravitational wave detectors can spot waves created by the mergers of neutron stars and black holes. As of the end of 2018, we've seen 10 mergers of black hole pairs and one merger of two neutron stars. As more sightings build up, astronomers will be able to see patterns in the numbers and masses of known black holes, which helps inform theories of how they form and change over time.

But we stand to learn even more from events that emit both gravitational waves and light. On August 17, 2017, astronomers got their first chance to see one of these events, when signals reached Earth from two merging neutron stars—ultra-dense leftovers from dead stars—that spiraled around each other and collided. The union not only released gravitational waves, it also triggered a visible explosion called a kilonova.

White dwarf stars that orbit each other, as in this artist's depiction, provide a realiable source of gravitational waves for scientists to study.

The new object that formed—most likely a black hole—fired a jet of high-speed particles through the surrounding haze, creating an afterglow that was visible for days to weeks afterward. This single event provided powerful evidence that colliding neutron stars probably make much of the universe's heavy elements, such as gold and silver. Like electronics and jewelry? Thank neutron stars.

Physicists were also able to use the detection to test Einstein's theory of relativity as never before. Relativity predicts that light and gravitational waves from the same event should travel through space the same way. Other theories of gravity, however, predict that the two should arrive at Earth at markedly different times. In the actual event, the light and gravitational waves arrived within seconds of each other—which means that gravitational waves and light react to obstacles in almost exactly the same way, within one part in a million billion.

Gravitational waves also help clarify other aspects of our universe's foundation. For instance, the Hubble Constant, a measure of how quickly the universe is expanding, has been tricky to pin down. Measurements of the early universe's afterglow yield one number, but estimates made using much younger stars yield another number. Is the discrepancy just a sampling issue or error? Or has the Hubble constant changed over time—suggesting the presence of new, bizarre particles and forces?

By acting as "standard sirens," gravitational wave detections provide an independent way to calculate the Hubble constant, making them the ultimate referee in this cosmic debate.

Feeling Gravitational Waves

The effect that gravitational waves have on Earth is thousands of times smaller than the width of a proton, one of the particles that makes up an atom's nucleus. That said, gravitational waves weaken the farther they travel, much like ripples on a pond. The closer you are to two merging black holes, the more you'd be stretched and strained.

But as trippy as it sounds, a gravitational wave stretches and compresses a given object as a percentage of the object's size. If Earth were as far from the black-hole merger that yielded LIGO's first detection as it is from the sun, gravitational waves would have stretched the planet by more than three feet. But peoples' bodies would be strained by just a millionth of a meter, far less than the compression you feel when you jump up and land on the ground.

Gravitational waves' proportional nature is why LIGO and other observatories have such large arms. The bigger the observatory, the bigger—and more detectable—the changes from a wave become.

Gravitational-wave Observatories

In 2017, the European observatory Virgo opened outside of Pisa, Italy, joining LIGO and Germany's GEO600 detector. And more such facilities are coming online: Japan's KAGRA detector, the first built underground, should be opening soon, and India is making plans to build its own detector.

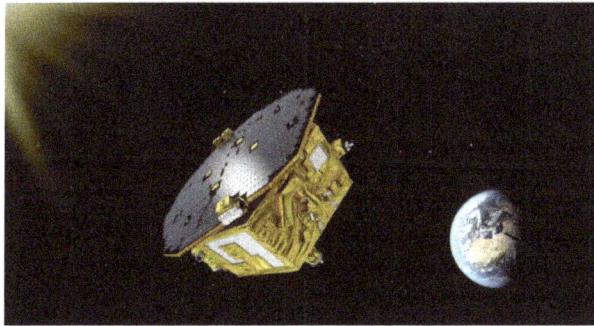

This artist's concept shows ESA's LISA Pathfinder spacecraft, which launched on a mission to detect gravitational waves.

In addition, there are plans to launch large, space-based observatories. The European Space Agency plans to put a detector called LISA into orbit around the sun in the 2030s. In 2015, the ESA launched the LISA Pathfinder spacecraft to test the necessary technology. Chinese researchers have proposed a similar space-based detector called TianQin.

Meanwhile, astronomers keep monitoring arrays of pulsars to track very low-frequency gravitational waves. The thinking is that as a wave sweeps through, it would temporarily alter the timing of each pulsar's rotation.

Gravitational Lensing

Most people are familiar with the tools of astronomy: telescopes, specialized instruments, and databases. Astronomers use those, plus some special techniques to observe distant objects. One of those techniques is called "gravitational lensing."

This method relies simply on the peculiar behavior of light as it passes near massive objects. The gravity of those regions, usually containing giant galaxies or galaxy clusters, magnifies light from very distant stars, galaxies, and quasars. Observations using gravitational lensing help astronomers explore objects that existed in the very earliest epochs of the universe. They also reveal the existence of planets around distant stars. In an uncanny way, they also unveil the distribution of dark matter that permeates the universe.

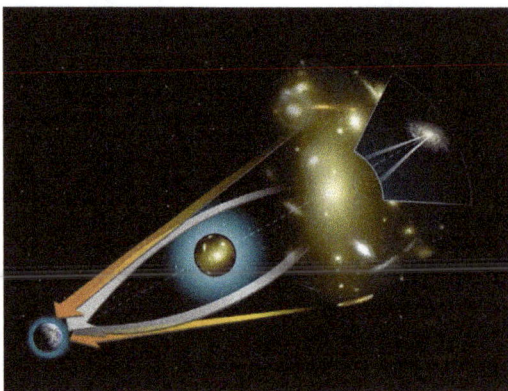

Gravitational lensing and how it works. Light from a distant object passes by a closer object with a strong gravitational pull. The light is bent and distorted and that creates "images" of the more distant object.

The Mechanics of a Gravitational Lens

The concept behind gravitational lensing is simple: everything in the universe has mass and that mass has a gravitational pull. If an object is massive enough, its strong gravitational pull will bend light as it passes by. A gravitational field of a very massive object, such as a planet, star, or galaxy, or galaxy cluster, or even a black hole, pulls more strongly at objects in nearby space. For example, when light rays from a more distant object pass by, they are caught up in the gravitational field, bent, and refocused. The refocused "image" is usually a distorted view of the more distant objects. In some extreme cases, entire background galaxies (for example) may end up distorted into long, skinny, banana-like shapes via the action of the gravitational lens.

The Prediction of Lensing

The idea of gravitational lensing was first suggested in Einstein's Theory of General Relativity. Around 1912, Einstein himself derived the math for how light is deflected as it passes through the Sun's gravitational field. His idea was subsequently tested during a total eclipse of the Sun in May 1919 by astronomers Arthur Eddington, Frank Dyson, and a team of observers stationed in cities across South America and Brazil. Their observations proved that gravitational lensing existed. While gravitational lensing has existed throughout history, it's fairly safe to say that it was first discovered in the early 1900s. Today, it is used to study many phenomena and objects in the distant universe. Stars and planets can cause gravitational lensing effects, although those are hard to detect. The gravitational fields of galaxies and galaxy clusters can produce more noticeable lensing effects. And, it now turns out that dark matter (which has a gravitational effect) also causes lensing.

Types of Gravitational Lensing

Now that astronomers can observe lensing across the universe, they've divided such phenomena into two types: strong lensing and weak lensing. Strong lensing is fairly easy to understand — if it can be seen with the human eye in an image, then it's strong. Weak lensing, on the other hand, is not detectable with the naked eye. Astronomers have to use special techniques to observe and analyze the process.

Due to the existence of dark matter, all distant galaxies are a tiny bit weak-lensed. Weak lensing is used to detect the amount of dark matter in a given direction in space. It's an incredibly useful tool for astronomers, helping them understand the distribution of dark matter in the cosmos. Strong lensing also allows them to see distant galaxies as they were in the distant past, which gives them a good idea of what conditions were like billions of years ago. It also magnifies the light from very distant objects, such as the earliest galaxies, and often gives astronomers an idea of the galaxies' activity back in their youth.

Another type of lensing called "microlensing" is usually caused by a star passing in front of another one, or against a more distant object. The shape of the object may not be distorted, as it is with stronger lensing, but the intensity of the light wavers. That tells astronomers that microlensing was likely involved. Interestingly, planets can also be involved in microlensing as they pass between us and their stars.

Gravitational lensing occurs to all wavelengths of light, from radio and infrared to visible and ultraviolet, which makes sense, since they're all part of the spectrum of electromagnetic radiation that bathes the universe.

The First Gravitational Lens

The pair of bright objects in the center of this image were once thought to be twin quasars.
They are actually two images of a very distant quasar being gravitationally lensed.

The first gravitational lens (other than the 1919 eclipse lensing experiment) was discovered in 1979 when astronomers looked at something dubbed the "Twin QSO". QSO is shorthand for "quasi-stellar object" or quasar. Originally, these astronomers thought this object might be a pair of quasar twins. After careful observations using the Kitt Peak National Observatory in Arizona, astronomers were able to figure out that there weren't two identical quasars (distant very active galaxies) near each other in space. Instead, they were actually two images of a more distant quasar that were produced as the quasar's light passed near a very massive gravity along the light's path of travel. That observation was made in optical light (visible light) and was later confirmed with radio observations using the Very Large Array in New Mexico.

Einstein Rings

A partial Einstein Ring known as the Horseshoe. It shows the light from a distant galaxy
being warped by the gravitational pull of a closer galaxy.

Since that time, many gravitationally lensed objects have been discovered. The most famous are Einstein rings, which are lensed objects whose light makes a "ring" around the lensing object. On the chance occasion when the distant source, the lensing object, and telescopes on Earth all line up, astronomers are able to see a ring of light. These are called "Einstein rings," named, of course, for the scientist whose work predicted the phenomenon of gravitational lensing.

Einstein's Famous Cross

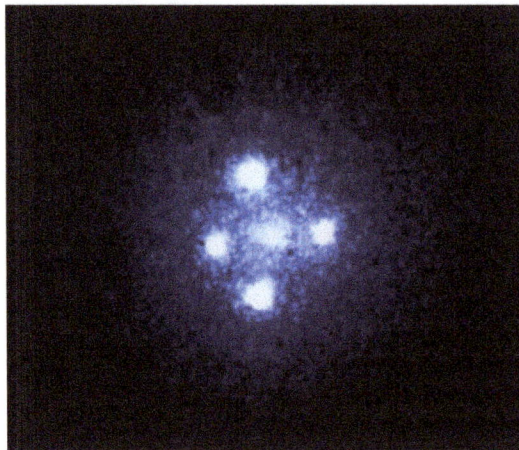

The Einstein Cross is actually four images of a single quasar (the image in the center is not visible to
the unaided eye). This image was taken with the Hubble Space Telescope's Faint Object Camera.
The object doing the lensing is called "Huchra's Lens" after the late astronomer John Huchra.

Another famous lensed object is a quasar called Q2237+030, or the Einstein Cross. When the light of a quasar some 8 billion light-years from Earth passed through an oblong-shaped galaxy, it created this odd shape. The lensing galaxy is much closer to Earth than the quasar, at a distance of about 400 million light-years. This object has been observed several times by the Hubble Space Telescope.

Strong Lensing of Distant Objects in the Cosmos

On a cosmic distance scale, Hubble Space Telescope regularly captures other images of gravitational lensing. In many of its views, distant galaxies are smeared into arcs. Astronomers use those shapes to determine the distribution of mass in the galaxy clusters doing the lensing or to figure out their distribution of dark matter. While those galaxies are generally too faint to be easily seen, gravitational lensing makes them visible, transmitting information across billions of light-years for astronomers to study.

This is Abell 370, and shows a collection of more distant objects being lensed by the combined gravitational pull of a foreground cluster of galaxies. The distant lensed galaxies are seen distorted, while the cluster galaxies appear fairly normal.

Astronomers continue to study the effects of lensing, particularly when black holes are involved. Their intense gravity also lenses light, as shown in this simulation using an HST image of the sky to demonstrate.

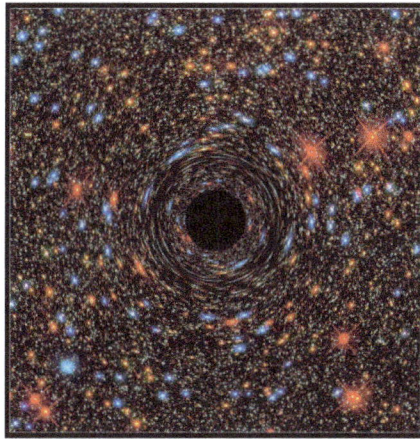

This computer-simulated image shows a supermassive black hole at the core of a galaxy. The black region in the center represents the black hole's event horizon, where no light can escape the massive object's gravitational grip. The black hole's powerful gravity distorts space around it like a funhouse mirror, in a process known as gravitational lensing. Light from background stars is stretched and smeared as the stars skim by the black hole.

References

- Black-holes: sciencealert.com, Retrieved 20 February, 2019

- Brill, Dieter (2012). "Black Hole Horizons and How They Begin". The Astronomical Review. 7 (1): 25–35. Bibcode:2012astrv...7a..25B. Doi:10.1080/21672857.2012.11519694. Retrieved 1 September 2012

- Singularity: universetoday.com, Retrieved 30 March, 2019

- Topper, David (2012). How Einstein Created Relativity out of Physics and Astronomy (illustrated ed.). Springer Science & Business Media. P. 118. ISBN 978-1-4614-4781-8. Extract of page 118

- Introduction-to-gravitational-lensing-4153504: thoughtco.com, Retrieved 1 April, 2019

- "First Successful Test of Einstein's General Relativity Near Supermassive Black Hole". Www.mpe.mpg.de. Retrieved 2018-07-28

- Gravitational-waves, reference, space, science: nationalgeographic.com, Retrieved 2 May, 2019

- Iorio, L. (August 2010). "Juno, the angular momentum of Jupiter and the Lense–Thirring effect". New Astronomy. 15 (6): 554–560. Arxiv:0812.1485. Bibcode:2010newa...15..554I. Doi:10.1016/j.newast.2010.01.004

- Sergei Kopeikin; Michael Efroimsky; George Kaplan (25 October 2011). Relativistic Celestial Mechanics of the Solar System. John Wiley & Sons. ISBN 978-3-527-63457-6

PERMISSIONS

All chapters in this book are published with permission under the Creative Commons Attribution Share Alike License or equivalent. Every chapter published in this book has been scrutinized by our experts. Their significance has been extensively debated. The topics covered herein carry significant information for a comprehensive understanding. They may even be implemented as practical applications or may be referred to as a beginning point for further studies.

We would like to thank the editorial team for lending their expertise to make the book truly unique. They have played a crucial role in the development of this book. Without their invaluable contributions this book wouldn't have been possible. They have made vital efforts to compile up to date information on the varied aspects of this subject to make this book a valuable addition to the collection of many professionals and students.

This book was conceptualized with the vision of imparting up-to-date and integrated information in this field. To ensure the same, a matchless editorial board was set up. Every individual on the board went through rigorous rounds of assessment to prove their worth. After which they invested a large part of their time researching and compiling the most relevant data for our readers.

The editorial board has been involved in producing this book since its inception. They have spent rigorous hours researching and exploring the diverse topics which have resulted in the successful publishing of this book. They have passed on their knowledge of decades through this book. To expedite this challenging task, the publisher supported the team at every step. A small team of assistant editors was also appointed to further simplify the editing procedure and attain best results for the readers.

Apart from the editorial board, the designing team has also invested a significant amount of their time in understanding the subject and creating the most relevant covers. They scrutinized every image to scout for the most suitable representation of the subject and create an appropriate cover for the book.

The publishing team has been an ardent support to the editorial, designing and production team. Their endless efforts to recruit the best for this project, has resulted in the accomplishment of this book. They are a veteran in the field of academics and their pool of knowledge is as vast as their experience in printing. Their expertise and guidance has proved useful at every step. Their uncompromising quality standards have made this book an exceptional effort. Their encouragement from time to time has been an inspiration for everyone.

The publisher and the editorial board hope that this book will prove to be a valuable piece of knowledge for students, practitioners and scholars across the globe.

INDEX

www.ingramcontent.com/pod-product-compliance
Lightning Source LLC
Chambersburg PA
CBHW061259190326
41458CB00011B/3714